T0390497

Star Bound

Outward Odyssey
A People's History of Spaceflight

Series editor
Colin Burgess

Star Bound

A Beginner's Guide to the
American Space Program, from Goddard's Rockets
to Goldilocks Planets and Everything in Between

Emily Carney and Bruce McCandless III

UNIVERSITY OF NEBRASKA PRESS • LINCOLN

© 2025 by Emily Carney and
Bruce McCandless III

Portions of this book have previously appeared in Bruce McCandless III, *Wonders All Around: The Incredible True Story of Astronaut Bruce McCandless II and the First Untethered Flight in Space* (Greenleaf Publishing Group, 2021).

"The Sister of Icarus" was originally published in *Rattle* 37 (Summer 2012).

All rights reserved
Manufactured in the
United States of America

The University of Nebraska Press is part of a land-grant institution with campuses and programs on the past, present, and future homelands of the Pawnee, Ponca, Otoe-Missouria, Omaha, Dakota, Lakota, Kaw, Cheyenne, and Arapaho Peoples, as well as those of the relocated Ho-Chunk, Sac and Fox, and Iowa Peoples.

Library of Congress Control Number: 2024013170

Set in Garamond Premier Pro
by Lacey Losh.

The Sister of Icarus

At the craft store, two angel wings
reveal themselves under the sheer
shirt of the girl in front of me.

Clear indigo lines etched over both
shoulder blades and beyond, each feather
meticulously outlined. I remember her

apple-white skin, chestnut hair, the sound
of coins clinking and her slipping away

swinging her bag of purchases:
feathers, glitter, and glue.

—Mary Ellen Redmond

Star Bound: The condition of dependence on a particular star, as life on Earth is dependent on the sun. Alternate meaning: The intention or decision to travel toward a star or stars outside our solar system, a journey that we will eventually take in order to establish human life elsewhere in the universe.

Contents

List of Illustrations ix

Acknowledgments xi

A Few Dates to Remember xiii

NASA's Crewed Spaceflights (So Far) xv

1. First Principles 1

2. The Wizard of Worcester 9

3. Rocketry and Death 19

4. The SS Major and the Suicide Squad 27

The All-Time Greatest
Space Exploration Playlist 39

5. A Starting Gun Called Sputnik 43

6. Shadows in the Sky 55

7. Project Mercury 65

Eleven Boffo Space Books
to Launch at Your Brain 75

8. Gemini's Forgotten Flights 77

9. The Rise and Fall of the
American Astronaut 85

The Best Space Stuff You Can Watch 99

10. Apollo and the First Man on the Moon 103

11. *Skylab* and the Renaissance
of American Science 119

NASA's Eleven Coolest Astronauts 131

12. Probes, Rovers, and the Golden Record 133

13. The Butterfly and the Bullet 141

NASA's Eleven Greatest Missions 159

14. Sleeping with the Russians 161

15. False Starts, Missteps,
and the Promise of Artemis. 169

America's Eleven Biggest Space Losses 177

16. The New Space Race 181

17. The Commercialization of Space 189

Eleven Everyday Benefits of the
American Space Program 209

18. Curse You, Gene Roddenberry! 213

The Eleven Most Persistent Rumors,
Riddles, and Conspiracy Theories
about the American Space Program. 219

19. Looking Outward 223

Eleven Fearless Space Predictions 237

20. Why It's Worth It. 239

Sources 251

Illustrations

Following page 132

1. Robert Goddard before launching
   his first liquid-fueled rocket, March 1926

2. Dr. James Pickering, Dr. James Van Allen,
   and Wernher von Braun holding replica
   of the Jupiter-C rocket

3. Scott Carpenter suiting up, May 1962

4. Chris Kraft, early 1960s

5. Jim Lovell and Frank Borman
   after *Gemini 7* space voyage

6. Dave Scott peering out of
   Apollo 9 command module

7. Buzz Aldrin saluting the American flag
   on the lunar surface, 20 July 1969

8. Al Shepard practicing for *Apollo 14*
   moonwalk with Ed Mitchell

9. NASA officials discussing *Skylab*'s
   missing micrometeoroid shield

10. American astronauts with
    Soviet cosmonauts, July 1975

11. Anna Fisher

12. Ron McNair, who filmed portions of STS-41B in February 1984

13. Carl Meade and Mark Lee testing the "rescue jetpack"

14. Eileen Collins

15. SpaceX Crew Dragon capsule, 2020

16. View of Pillars of Creation taken by Hubble Space Telescope, 2014

17. Crew of *Artemis 2*

18. Saturn's moon Enceladus

Acknowledgments

The authors thank Francis French, Dave Bohlmann, John Whisenhunt, Jay Gallentine, Angela Joyce, Chris Kelly, Greg Bollendonk, Lois Huneycutt, Michael Bear, Dave Shomper, Brian Dunning, Don Erwin, Roger Worthington, David Worthington, Bruce McCreary, Kevin H. Spencer, and Johannes Kemppanen for their help in telling this story. We couldn't have done the math parts, the grammar parts—heck, all the difficult, you know, *content* parts—without you. The mistakes are our fault, not yours. Lord knows you tried.

No words suffice to express our gratitude to our long-suffering spouses, Steve Carney and Pati Fuller McCandless, so we will just pick three at random: *nozzle*, *optimal*, and *atlatl*.

We also admire and wish to acknowledge Gary Minar, Lou Ramon, Katherine Johnson, Dee O'Hara, Ed White, Dick Smith, Bill Tindall, Nancy Grace Roman, Ron Sheffield, Bobbie Johnson, Suzanne Dodd, Bill Moon, Franklin Chang-Díaz, Todd Bailey, Jefferson Hall, Dr. Joe Kerwin, Ed Rezac, Dr. Don Robbins, Dr. Stu Nachtwey, Dr. Ron McNair, David McCann, Poppy Northcutt, Yosio Nakamura, Ed Whitsett, Judith Love Cohen, Dave Schultz, Dr. Ed Lu, Judy Resnik, Bill Bollendonk, and the thousands of other brilliant, dedicated people who actually created the history that we're only writing about.

And finally, we shout out additional thanks and all good wishes to the Space Hipsters Facebook group—as fun, curious, mildly obsessive, and altogether excellent a family of space nerds as ever existed in any galaxy anywhere . . . that we know of . . . so far.

Onward!

A Few Dates to Remember

25 July 1865: Jules Verne publishes his novel *From the Earth to the Moon*, which inspires a number of brilliant, somewhat peculiar individuals to start dreaming about space travel.

26 March 1926: Robert Goddard launches the first liquid-propelled rocket.

31 October 1936: Frank Malina, Jack Parsons, and the Suicide Squad fire up their first rocket motor experiment in Pasadena, California.

20 September 1945: The first German rocket engineers, including Werner von Braun, arrive in America. Their arrival is kept secret.

4 October 1957: The Soviet Union puts the first satellite (Sputnik) in orbit. The Soviets launch the much larger Sputnik 2 shortly thereafter.

31 January 1958: America deploys *its* first satellite, Explorer 1.

12 April 1961: Cosmonaut Yuri Gagarin becomes the first person in space.

5 May 1961: Astronaut Alan Shepard becomes the first American in space.

18 March 1965: Alexei Leonov of the USSR makes the first "space walk."

3 June 1965: Ed White makes the first American space walk.

20 July 1969: *Apollo 11* astronauts Neil Armstrong and Buzz Aldrin walk on the moon.

20 August 1977: NASA launches the first of its two *Voyager* probes.

12 April 1981: The first space shuttle leaves Earth.

25 April 1990: Shuttle astronauts deploy the Hubble Space Telescope, hoping it will have at least a fifteen-year lifespan. At publication time, the telescope was still functioning well into its thirty-fourth year.

8 July 2011: The last space shuttle flight, STS-135, takes off.

31 May 2020: SpaceX's Crew Dragon capsule restores America's ability to fly astronauts to the International Space Station.

26 September 2022: NASA's DART mission alters the orbit of a binary asteroid—a spectacular first step in figuring out how to avert a potentially disastrous future meteor impact with Earth.

NASA's Crewed Spaceflights (So Far)

Mercury (1961–63, 6 missions)

X-15 (1962–63, 13 spaceflights)

Gemini (1964–66, 10 missions)

Apollo (1967–72, 11 missions)

Skylab (1973–74, 3 missions)

Apollo-Soyuz Test Project (1975, 1 mission)

Space Shuttle (1981–2011, 135 missions)

International Space Station (2000–present, 71 expeditions)

SpaceX Crew Dragon (2020–present, 9 missions)

Boeing Starliner (2024–present, 1 mission [incomplete])

Star Bound

1

First Principles

How to find space—and why it's so difficult to get there.

Soon four people will feel the earth break beneath them. They'll sweat tiny diamonds. Their stomachs will churn with exhilaration and dread as the most powerful rocket the American space agency has ever built lifts them from a Florida launchpad to begin the quarter-million-mile trip to the moon. Their journey will jump-start what NASA calls the Artemis program. On the lunar surface, these astronauts will create humanity's first long-term settlement away from Earth, a tiny first step on what is sure to be an arduous, multi-millennia journey into the cosmos.

So why is it taking so long?

Neil Armstrong set foot on the moon half a century ago. Since then, men and women from numerous nations have spent years of combined time in space labs and shuttles, Russian rockets and Chinese capsules. The International Space Station has been orbiting Earth for over two decades. American probes have visited every planet in our solar system, many of its moons, and several of its asteroids. NASA is planning—and *planning*—the first human expedition to Mars. Given all these accomplishments, it may seem odd that we're just now talking about a return to the moon and a first attempt at life on a rock other than our own.

But this is mostly a matter of perspective. Imagine Earth's entire existence—from the beginning of the solar system to the present—as a typical twenty-four-hour day. The very first single-celled organisms show up at around 4 a.m. Life thereafter evolves in fits and starts (mostly fits) with algae appearing early in the afternoon and sexual reproduction starting up just after 6 p.m., as it still does in parts of Scandinavia. Jellyfish shimmy into view after dinner. Land plants arrive toward the end of prime time, and dinosaurs finally show up around the hour most of us are heading to bed. You get the idea. We've been here for a heartbeat. Humankind finally stumbles onto the historical stage a minute before midnight, scratching and squabbling and worrying about its receding hairline, which is just now saying goodbye to its eyebrows. Most of these early hominids are busy digging for grubs. A man's gotta eat, after all. But one of the bunch can't help herself. It's nighttime, remember, and she's fascinated by the lights she sees overhead, glittering like the play of sunshine on a distant dark sea. She calls her cousins over to share the view. One of them, not the smartest, perhaps, raises a hand, attempting to touch the luminous objects that seem so close . . .

Now we're going back to the moon. It's been several decades, true, but in archeological terms, we are still moving at breakneck speed. Which is good, because our exploration of the universe (or "space," which is another way of saying the same thing) has barely begun. It turns out that space is big—"vastly, hugely, mind-bogglingly big," in the words of novelist Douglas Adams. "Space is to place," the French essayist Joseph Joubert famously posited, "as eternity is to time." Think about *that* statement for a minute. The universe is so large that not even light, which travels faster than anything we know of, can make it all the way across. This is because space is not only big. It's getting bigger.

Here's another analogy. If our spacefaring species were living in the sixteenth century rather than the twenty-first and setting out by sail to explore the world rather than riding rockets into the solar system, we would still be inside the harbor. In fact, we could still jump from the ship and land on the dock. Even this tiny bit of progress has cost us. We have lost lives in our halting attempts to explore the heavens. We will doubtless lose more. But we're doing it anyway—and none of our past adventures will be as difficult, dangerous, or miraculous as the one to come.

If you're just now tuning into this effort, you're in luck. It's about to get good.

Our Aim

Star Bound is the story of how we got to where we are today, with some guesses as to where we're going next. It's an introductory text. It reeks of death and disaster. It's intended for the general reader, assuming the general reader is a little like us: curious but not mechanically minded, intrigued by the saga of our first forays into space without knowing the mind-grinding physics behind it all. Our narrative comes complete with bias, inappropriate emphasis, and all the other shortcomings of authorial discretion. You, the reader, could find all of the information contained in this book in other sources—in some cases, many other sources.

But information without organization can be frustrating rather than illuminating. You might also find it tedious to read hyper-detailed data presented by authors who are technically literate and frighteningly intelligent. Rest assured that you will encounter no such indignities in these pages. What we've tried to do in *Star Bound* is present a coherent, if simplified, story. We hope you'll learn a hundred things. We hope you'll pester friends and relatives with a golden nugget you first beheld in these pages. And we hope our tale is readable in the space of around three hours—the duration of a flight from, say, Austin, Texas, to Washington DC.

If you'd like to check our sources, we can keep you busy all the way to Newark.

Defining Space

Star Bound is not an engineering text.

And that's okay. One of the biggest hurdles to engagement with the American space program is other space enthusiasts, who can be territorial and high-handed, like self-anointed priests of a technological cult. Fact is, there's no entrance exam for enjoying the Apollo saga or for following NASA's plans to visit Mars. The story of space exploration is weirder and more compelling than you've been led to believe—a map of missed opportunities, phony promises, heart-stopping accidents, and astonishing achievements. It's a human story, and because this is so, it's fascinating beyond reason and beautiful beyond analysis. Don't let the gatekeepers distract you.

Nevertheless, it's important to understand some of the challenges involved in sending a person into space. Here, perhaps, is the first: *What is "space"?* There are many ways to answer this question. But for practical rather than philosophical purposes, there are two definitions, both of which involve altitude—that is, one's distance above our planet. According to the U.S. Air Force, "space" begins at fifty miles above sea level on Earth. Another definition states that space starts at one hundred kilometers (approximately sixty-two miles) above the planet. This hundred-kilometer point is called the Kármán line, in honor of engineer and physicist Theodore von Kármán, the man who proposed it. Kármán's is the more widely accepted delineation of where space begins, so we'll use it in *Star Bound*. It's important to have such a definition because the skies up to the boundaries of "space" can be claimed, and policed, by the countries beneath them. But above this point, space is open to travel by all. You can thank the old Soviet Union for that, as we shall see.

Not everyone agrees that either the air force measure or the Kármán line is the right boundary. Atmospheric conditions rarely arrive neatly packaged, in the way that a hundred-kilometer mark, or "line," might suggest. The matter of where space begins is less a number than a condition. It's the altitude at which Earth's atmosphere has dissipated to close to nothing. Because there is so little atmosphere, there is little atmospheric "drag," or friction, on an object traveling at this altitude. This means that such an object can no longer take advantage of the differences in air pressure between the underside of its wings and the upper side, as an airplane does, to fly. Above the Kármán line, which is the delineation we'll use, objects travel in a way determined by orbital dynamics, the interplay of velocity, distance, and gravitational pull—specifically, in most of what we will be studying, the gravitational pull of Earth.

Sixty-two miles is not that far. If your car could travel straight upward, you could drive to space in an hour—or roughly seventeen Taylor Swift songs. But even here, not so high above the planet, the typical aspects of space are present. The gentle curve of Earth's horizon is clearly evident. There's not enough oxygen at this altitude to sustain human life. (This actually becomes true at a measly five miles above sea level, as anyone who has climbed Mt. Everest can attest.) It's cold up here, but atmospheric pressure is nil, which means that your blood would literally boil in your veins if you left your spaceship without a pressure suit. The sky at this altitude is black, not blue, because there's not enough atmosphere to diffuse the sun's rays in such a way that we see more blue than, say, red, a phenomenon that occurs as a result of something called Rayleigh scattering. Also, people float. You may say that this is because there is no gravity to keep a person in place, but this is not true. At sixty-two miles up, Earth's gravitational pull is still around 97 percent of what it is on the planet's surface. In fact, weightlessness on the International Space Station or any other spacecraft in Earth orbit is the result not of reduced gravity but of another phenomenon altogether. We'll talk about this later, when you are in a better mood.

Getting There: Atmosphere, Mass, and Gravity

According to the most common definition, then, getting to space means getting at least sixty-two miles above Earth. This is hard to do, for a couple of reasons. First, our atmosphere is a soupy mixture of gases like oxygen, nitrogen, hydrogen, and argon. While our atmosphere is nowhere near as thick as that of, say, Venus, it is dense enough to create resistance to an object moving through it. Our air is solid enough for birds to glide on. It is thick enough that the friction caused by an object moving through it at high speed can generate tremendous heat. In some cases, as with the meteors that bombard us from outer space, our atmospheric incinerator is helpful. It burns up most of the projectiles the cosmos throws at us before they ever make it to the ground. For astronauts seeking to return to the planet, though, the heat is not helpful at all. Indeed, reentry into Earth's atmosphere is a phase of spaceflight that occasions considerable dread. But atmospheric drag increases the difficulty of leaving Earth as well.

Second, and more significantly, anything we try to send upward—from a baseball we throw from center field to a Falcon 9 rocket we launch from Kennedy Space Center—has mass, and mass means inclusion in the great universal dance of gravity. No one knows what gravity is, in a qualitative sense. Even Sir Isaac Newton, who formulated the laws of its behavior, declined to speculate in this regard—and he was an alchemist, for Pete's sake. Nevertheless, we know that gravity exists. It is a constant, a condition of existence, a blessing and a curse. It is attraction—invisible, inescapable, and non-negotiable. Brian Clegg, writing in 2021 for the BBC's *Science Focus* magazine, says, "It's a property of matter, of stuff. In a nutshell: all matter is attracted to all other matter. The more matter there is, and the closer objects are to each other, the bigger that attractive force." The author Mary Roach, in her book *Packing for Mars*, offers this: "Gravity is the pull, measurable and predictable, that one mass exerts on another. . . . Gravity is why there are suns and planets in the first place. It is practically God."

It's tempting, but not entirely accurate, to speak of gravity as a "force." Gravity is our way of describing the fact that mass seems to warp, or bend, the space around it. Mass also warps *time*, as Albert Einstein told us, but that's a discussion for a much trippier book. The bigger the mass, the bigger the bend. Objects aren't so much pulled toward each other as they *fall* toward each other. We don't know why gravity's effects occur, but we do know pretty precisely *how* they occur. A ball is thrown straight up, and it comes straight down. An arrow is shot at a forty-five-degree angle to the earth. The arrow travels up and out until the energy imparted to it is spent, and then it falls back to the ground. We can measure gravity's effects. More importantly, we can *predict* gravity's effects. We do so every day, whether we're shooting baskets or jumping over a puddle or planning to send a rocket to one of Saturn's moons.

For purposes of discussion, try this. Throw a five-pound rock in the air—straight up, if you can. We suggest holding the stone in both hands and then boosting it up, as you'd boost the foot of someone desperately attempting to escape from a foreign jail or detention center through a high and awkwardly placed window. (Don't ask us how we know this.) If you're robust and athletic, as our readers tend to be, you may be able to propel the rock fifty feet. If so, congratulations. But consider something. You've just sent your projectile a mere 1/6,547 of the sixty-two miles into space. That projectile, by the way, is only 1/1,240,000 of the weight of a fully fueled, 6.2-million-pound moon rocket.

Space, as the engineers like to say, is *hard*.

Rockets!

So how do we overcome gravity to get a person up through the atmosphere of Earth? As mentioned above, an airplane uses differentials in air pressure to function. Air flowing over the top of a wing moves faster than air flowing along the underside; the relative difference in pressure between lower and upper surface creates *lift*. An airplane can use this principle to climb to an altitude that qualifies as space, but it can't sustain such altitudes. As air pressure disappears, the plane's wings can no longer generate lift—so this option won't work.

A helium balloon can ascend to a height of around twenty miles. It climbs because the helium in the balloon is less dense than the surrounding atmosphere. However, at around twenty miles, the air thins to the point where the helium inside the balloon and the atmosphere outside it are approximately the same density. No ascent occurs after this.

Some scientists have proposed the use of a "space elevator" to get people and materials from Earth to space and back again. This device is generally conceived of as a tether or cable fastened to Earth at a point along the equator and reaching thousands of miles above the planet, with the tether held in tension by the competing forces of gravity and centrifugal force. Transportation mechanisms (sometimes called "crawlers") could then be attached to the cable and moved up or down. While the idea seems promising, it is, as yet, impractical. We just don't have a material strong enough to withstand the stresses that would be put on the elevator's tether.

We could also, in theory, shoot ourselves into space from a giant gun, as at least one science fiction writer has posited and as many Americans think about doing immediately after watching political commercials. This might work on the moon, which has very little atmosphere and—at its surface—only around 16 percent of the gravitational force experienced on the surface of Earth. But so far no one has figured out how a human being could survive the heat and sudden, much more violent, effects of acceleration that would be involved in employing such a method here on our planet.

Because we can't fly, float, or climb on a suspended tether, we need a different method to get into and maneuver in space. The method we have used so far is rocket propulsion. It is, to put it simply, the use of controlled explosions—the rapid combustion of chemical fuels, channeled in such a way as to propel an object away from the direction of the blast. With a large enough rocket, controlling a powerful enough explosion, space travel becomes possible. We learned this by means of many untimely explosions along the way.

And this brings us to Robert H. Goddard, the bespectacled god of Really Big Ballistics. *Ballistics* is one of those words that just sounds cool. We say it often at social gatherings, and we are confident that someday someone will ask us why. It's the study of how projectiles—iron cannonballs, *human* cannonballs, nuclear missiles, suborbital spacecraft—move in flight, from point A (launch) to point B (landing, or impact).

A child of the nineteenth century, Goddard was a loner, suspicious of outsiders, aware that his ideas marked him as a crank to most of his fellow Americans. His life was full of accomplishments, but one in particular stands out. A famous photograph sets the scene. Goddard is standing on a frigid farm in Auburn, Massachusetts, next to what looks like playground equipment—a primitive jungle gym, perhaps. The reclusive tinkerer with the bottle-brush mustache has no conception of what a "global positioning system" might be. It hasn't even been imagined yet. He can't watch televised images from across the Atlantic Ocean in real time. He has no idea that huge oceans lie beneath the ice of Jupiter's moon, Europa. Nevertheless, he's proud of his, er, *jungle gym*. This is because in reality, the metal-tubed contraption is a liquid-fueled rocket, the first of its kind, that Goddard calls "Nell," and he's prepared to ignite the explosion that will launch it. The sky is gun-barrel gray. Someone's goat has gotten loose in the field across the creek, but no matter. It's a chilly afternoon in March of 1926, and the world is about to change. For better. For worse.

And forever.

2

The Wizard of Worcester

America's home-grown Daedalus was shy and bald and dreamed of machines that could sail through space. He was widely mocked as a result. But it turns out he was right.

To be sure, there were rockets around long before Robert Goddard was born. Chinese artisans learned well over a thousand years ago how to produce the explosive mixture of potassium nitrate, sulfur, and charcoal we call gunpowder and that they referred to as *huo yao*, or the "fire drug." They realized that the energy released by gunpowder during combustion could, if properly channeled, send a small object skyward. The result was fireworks, which were first documented in China during the illustrious Song dynasty. By way of the Middle East, the ancient world's open-air market for silk, spice, and secrets, pyrotechnics reached Europe by the end of the thirteenth century.

As spectacular and celebratory as fireworks can be, gunpowder was eventually employed for less pleasant work as well. Unsurprisingly, the first gunpowder weapons—rockets and "hand cannons" mounted on spears—were also developed in China. They were crude by modern standards but relatively advanced for an era when European knights were still flailing around in the mud and whacking each other over the head with spiked balls and hammers. The first rockets were essentially explosive charges lashed to arrow shafts. They were similar in spirit to what every kid knows today as the dread of all dachshunds, the bottle rocket, with one end of the charge capped and the other left open to permit the venting of exhaust that propelled the charge and shaft toward the enemy. Such "fire arrows"—larger and more potent, of course—were used by Chinese soldiers of the Jin dynasty against a Mongol invasion force at the Siege of Kai-Feng in 1232–33. The weapons weren't decisive. The Mongols won, and the Jin dynasty crumbled not long afterward. But the fire arrows made an impression. The Mongols copied them and used them in future battles against their foes—including the neighboring Song dynasty, which was next to fall as the Mongols consolidated their hold on eastern Asia.

Meanwhile, fascinated by gunpowder's destructive potential, military engineers set about refining its use to create explosions that could push projectiles at high velocity down a rigid tube toward a target. This led to the development of metal-barreled cannons, muskets, and rifles, mainstays of human misery for hundreds of years now.

Rockets and barreled weapons like cannons work differently. Cannons use gunpowder or another type of explosive substance to fire projectiles. Rockets carry their own fuel—though this fuel was also, for many centuries, gunpowder—to launch and propel *themselves*. Basically, a rocket is an object that flies by burning fuel it carries within and directing the exhaust in one direction in order to push the rocket in the other direction. It is, as one early newspaper account put it, a "flying inferno."

The rocket's operation nicely illustrates Sir Isaac Newton's third law of motion, which states that for every action, there is an equal and opposite reaction. Hold an inflated balloon in your hand. Pinch the blowhole shut with your thumb and index finger. On the count of three, release your grip. Air will rush out of the balloon in one direction, while the balloon moves in the other. Just so, when a spaceship is launched from Kennedy Space Center, exhaust produced by the combustion of fuel and oxygen within the rocket goes down, and the rocket (ideally) goes up. The amount of exhaust, and the velocity at which it is expelled, determine the force with which the rocket will be driven skyward. This force is called *thrust*.

Rockets aren't built like balloons, of course. We tend to think of rockets as sleek, cylindrical, *pointy* things—and for good reason. That's what they typically are. Their arrow-like design decreases the amount of air a rocket has to push against as it flies. But if it weren't for the fact that the rocket is leaving from Earth, having to battle the resistance offered by the planet's atmosphere, a rocket could be any shape. Indeed, vehicles designed to be launched in the airless vacuum of space, like NASA's bat-like asteroid probe, *Lucy*, or the spidery lunar lander of the Apollo missions, take all kinds of strange forms. And structures built in space to remain in a more or less static orbit, like the International Space Station, resemble nothing so much as Tinker Toys assembled by a four-year-old between bites of carpet fiber and mashed potato.

Two Legs between Six

By the fifteenth century, armies in Asia, Europe, and the Middle East were using increasingly powerful gunpowder weapons. The extraordinary Joan of Arc and her French army faced Burgundian cannons at the Siege of Compiegne in the fifteenth century. Ottoman sultan Mehmet II conquered Constantinople in 1453 largely through the use of artillery, including massive bronze cannons with barrels that could be split apart and then reassembled for transport. Four hundred years later, the Union Army brought to bear some 3,325 artillery pieces—Napoleons, Howitzers, three-inch ordnance rifles—in its combat operations against the gray-clad legions of the Confederacy.

And during the First World War, the German armaments manufacturer Krupps produced several 258-ton cannons known as Paris guns that were capable of firing explosive projectiles some seventy-five miles. This was an extraordinary range, and many experts flatly disbelieved early reports of the weapons, theorizing that damage from the guns had been caused by aerial bombs instead. ("EXPERTS DUMBFOUNDED" read the headline of one contemporary newspaper account.) The projectiles fired from these weapons rose so high that artillery men had to take Earth's rotation into account when aiming their fire. Indeed, the Paris guns were so intimidating and destructive that Germany was expressly prohibited from possessing them by the terms of the agreements that ended the First World War. Germany complied with the ban. In the meantime, though, it started searching for some other way to terrify its neighbors.

While cannons and other barreled weapons grew steadily larger and more accurate, rocket technology advanced at a comparatively glacial pace. British military forces suffered heavy casualties in the eighteenth century when they encountered opponents in India who used rockets not much different from those constructed by the Chinese many centuries earlier. British inventor Sir William Congreve obtained samples of the weapons and set about improving their range, accuracy, and destructive capabilities. For example, Indian forces had tied their explosive charges to the side of the long bamboo rods that stabilized the rockets in flight; Congreve ran his wooden stabilizing rod up through the middle of his charge, which helped his rockets fly straighter. The British incorporated the weapons into their own methods of warfare, as when they used so-called Congreve rockets to attack America's Fort McHenry during the War of 1812.

Despite Congreve's work, rockets in those days were frightening and potentially lethal but still erratic. Their best use (or worst, depending on one's perspective) was when they were launched in significant numbers at a large target that couldn't get up and run away—the city of Copenhagen, for example, which the British burned to the ground with Congreve rockets in September of 1807. All in all, most military thinkers considered artillery a better battlefield alternative, and rocketry faded in importance again as the nineteenth century proceeded. The cannon was king. Indeed, such were the advancements in artillery engineering that French science fiction writer Jules Verne, in his 1865 novel *From the Earth to the Moon*, imagined a "space-ship" (as a contemporary newspaper coined the term) flying to the moon after being fired from a six-hundred-foot-long iron cannon called the Columbiad located in Tampa, Florida.

Verne got the launch site more or less right, even if he was wrong about how spacecraft would eventually leave the state, and *From the Earth to the Moon* ignited the passions of some of the great rocketry pioneers in the United States and Europe. As grand as his visions were, however, the author was also careful to warn about the dangers involved in attempts to harness the destructive power of propulsive explosions. In the novel, a grand wager regarding the ability of men to fly to Earth's satellite is made within the august confines of an organization called the Baltimore Gun Club, a favorite haunt of American artillery veterans. "The Yankees," Verne notes, with grudging admiration,

are engineers . . . by right of birth. Nothing is more natural, therefore, than to perceive them applying their audacious ingenuity to the science of gunnery. . . . It is but fair to add that these Yankees, brave as they have ever proved themselves to be, did not confine themselves to theories and formulas, but that they paid heavily, in *propria persona*, for their inventions. . . . Many had found their rest on the field of battle whose names figured in the Book of Honor of the Baltimore Gun Club; and of those who made good their return the greater proportion bore the marks of their indisputable valor. Crutches, wooden legs, artificial arms, steel hooks, caoutchouc jaws, silver craniums, [and] platinum noses, were all to be found in the collection; and it was calculated by the great statistician Pitcairn that throughout the Gun Club there was not quite one arm between four persons, and exactly two legs between six.

Deep Fantastic Blue

Born on 5 October 1882 in Worcester, Massachusetts, Robert Hutchings Goddard was a frail child, often ill, fussed over by an adoring grandmother. The son of a machinist father, he showed an early fondness for building gadgets, and he seems to have been interested in just about everything. He set off firecrackers and Roman candles, did rudimentary experiments with electricity, and, according to biographer David Clary, organized a team of neighborhood youths to dig a tunnel to China, which was apparently never completed. Young Goddard loved books and read Verne's *From the Earth to the Moon* and H. G. Wells's Martian invasion thriller *The War of the Worlds* "many times," says Clary. He had an epiphany in 1899 when, as a teenager, he climbed to the top of a cherry tree and became fascinated by the arc and expanse of the emptiness overhead. Other people have seen visions in the sky. The emperor Constantine reportedly saw a fiery cross in the heavens before an important battle in 312 CE. His forces won the battle, and he converted to Christianity as a result. But for Goddard, the vision wasn't in the sky. It *was* the sky. He began to wonder if human beings could travel beyond this azure barrier into the deep fantastic blue and possibly to other planets—specifically Mars, which many Americans believed at the time might be inhabited. "I was a different boy when I descended the tree from when I ascended," Goddard later wrote. "Existence at last seemed very purposive."

Purposive indeed. Young Goddard never ceased thinking about the possibility of space travel. He eventually became convinced that the way into the cosmos was through rockets. He spent most of his adult life—an industrious, practical, and occasionally paranoid life—building the metal beasts. As a young man, he earned a PhD in physics from Clark University, did research work at Princeton, and eventually returned to Worcester to join the faculty at Clark. In addition to his intellectual accomplishments, he was also a tinkerer, unafraid of a skinned knuckle or two, comfortable though not particularly skillful with a blowtorch and a crescent wrench. He may not have been the first person on the planet to see rockets as something other than dangerous toys or obsolete weapons, but he was apparently the first to start the greasy, tedious, occasionally dangerous work of constructing what he saw as the vehicle of the future. He seems to have dismissed early on the viability of both cannons and balloons for space travel, but he built numerous rocket prototypes to test his ideas. Most of these early rockets died inglorious deaths. They blew up. They fell down. They smashed themselves into the earth. Nevertheless, Goddard persisted. We remember him not because he dreamed but because he built. And *rebuilt*. And rebuilt again. His rockets gradually got bigger, flew higher, and traveled more or less where they were pointed.

He was also careful to patent his work, which gives us a handy paper trail of his accomplishments. It's a long trail, as he eventually earned 214 patents, some of which were awarded posthumously. On 1 October 1913, for example, Goddard filed a patent application for the "multistage rocket." Simply put, having multiple *stages*, a fancy rocket-guy word for "sections," means that a spacecraft can get rid of a section when the fuel in that section is depleted and the empty stage represents only unnecessary weight. In an enterprise where every pound counts, it's a key innovation that has helped to make orbital spaceflight the more or less commonplace proposition it is today.

On 15 May 1914, Goddard filed his next patent application for a high-altitude rocket powered by a liquid—as opposed to solid, like gunpowder—propelled rocket engine using gasoline as a fuel and liquid nitrous oxide as the oxidant. This was another important development, as we shall see. Because liquid propellants were expensive and relatively difficult to come by, though, Goddard continued to power his rockets with gunpowder and nitrocellulose smokeless powder. Using newly developed hardware called de Laval nozzles to channel the propulsive exhaust, he steadily increased the efficiency and thrust of his creations.

A note about liquid propellants. On Earth or in space, combustion requires oxygen. A rocket engine doesn't consume oxygen from the atmosphere. Rather, it carries its own, which it mixes with a flammable substance like hydrogen, gasoline, or kerosene to create a controlled, intense, and "aimable" explosion in order to move. But why, the reader may ask, a *liquid* fuel? And what is the significance of using liquid oxygen? Why this pale blue fluid, with paramagnetic properties, created by cooling oxygen down to ridiculous temperatures like −297 degrees Fahrenheit, which is only slightly warmer than the surface of Pluto? The answer is simple: liquefying oxygen or fuel makes it compact, and therefore easier, cheaper, and lighter for a rocket to carry—not because the substance is lighter, but because the tank containing it can be smaller. Consumption of liquid fuel and its oxidant is also controllable. It can be increased or decreased, and even *stopped*, whereas combustion of a solid fuel like gunpowder tends to be an all-or-nothing proposition; once it starts, it continues until the fuel is depleted. Liquid propellants have therefore been the primary drivers of rockets from Goddard's time right down to the present.

Based on the success of his early work, in January 1917 Goddard received a grant of $5,000 over five years from the Smithsonian Institution to fund his research. It was a substantial sum, and it allowed him to upgrade his hardware and propellant options. He worked alone when he could, since few people understood his ideas and even fewer agreed with them. Rocketry wasn't considered a legitimate academic pursuit in those days. It was more like a bad habit, similar to setting grass fires or playing the banjo. Nevertheless, the Smithsonian published Goddard's thoughts in a 1919 monograph called *A Method of Reaching Extreme Altitudes*. The work, which contained notes on Goddard's experiments with rocket engines and nozzle technology, was eagerly read by rocket enthusiasts abroad, including two Germans named Hermann Oberth and Wernher von Braun. The publication was less well received at home. The *New York Times* got wind of his proposal to send a rocket to the moon and ridiculed the professor in print. Summoning that peculiar blend of smugness and scientific error available only to newspaper editors, the *Times* announced that Goddard's plan was foolish because a rocket in space wouldn't have any atmosphere for its exhaust to push against. *Nothing to push against, dammit!* It was an embarrassing error on Goddard's part, the newspaper said, indicating that the good doctor lacked basic information like that "ladled out daily in high schools."

Nevertheless, on 16 March 1926, after years of trial, error, and a fair amount of exasperated head-shaking from friends and colleagues, Goddard was able to launch a liquid-fueled rocket forty-one feet into the air. Powered by a mixture of gasoline and liquid oxygen, the flight lasted 2.5 seconds and ended in a cabbage field 184 feet away from the launch site. By modern standards, it was a modest trajectory. The nearby Boston Red Sox baseball club fielded a sorry collection of misfits that season, a squad that included ham-and-eggers like Dud Lee, Boob Fowler, and Baby Doll Jackson. It was an organization still reeling from the colossal stupidity of selling a pug-nosed pitcher named Babe Ruth to the New York Yankees for a few bucks and a hot dog. Still, any player on the Sox could have thrown a baseball just as far, and considerably higher, than Goddard's rocket. Most could probably have thrown Goddard's *rocket* just as far. But the point was confirmation. Goddard had proven that a liquid-fueled rocket could work. The launch site is now a national historic landmark.

In July 1969, as *Apollo 11* neared the moon, the *New York Times* issued a formal retraction of its unkind words about Professor Goddard's work. Apparently, rockets *could* fly in space. Who knew? The newspaper regretted the error.

Summing Up

Goddard contracted tuberculosis as a young man. Though he survived the ordeal, he bore traces of the disease for the rest of his life. Tall, bald, and stoop-shouldered, with a reedy voice and a gentle sense of humor, he was the prototypical absent-minded professor, beloved by family, friends, and many of his colleagues. The rest of the nation wasn't quite sure what to think of him and his "moon rockets," as the popular press delighted in calling Goddard's temperamental metal canisters. Some revered him. Charles Lindbergh, for example, became a lifelong supporter. Others dismissed Goddard's work as fanciful nonsense. "Oh yes," Ray Bradbury wrote of his own childhood in the 1920s, "later on we were to remember that there were a few wild men like Professor Goddard stirring about. But no one gave him mind. He was a blathering idiot, a fool, a nothing." The "no one" here is hyperbole. In fact, Goddard was fussed over all his life by women who loved him. Perhaps as a result he was self-directed and convinced of his own correctness. On the other hand, he was also a little more satisfied with his own ideas and ways of doing things than was completely productive. He could be stubborn and unsystematic, working on a second set of problems before he'd solved the first. In his later years, experimenting in New Mexico with the aid of funding from the Daniel and Florence Guggenheim Foundation, Goddard became increasingly secretive, worried that his ideas were being pirated by nosy militarists in Germany and admiring young amateurs at home. He was not particularly helpful to either group—to his credit in the first instance but hardly praiseworthy in the second.

Certainly he could be difficult, self-centered, occasionally distrustful. In these regards he was a peculiarly American figure, a constructive contrarian who, like Edison and the Wright brothers, believed he could rewrite the laws of possibility with the help of the right collection of rivets and wires and a suitable energy source. His insistence on following his own path eventually resulted in his work being eclipsed by younger engineers, with larger, more specialized teams and stronger financial backing. Yet Goddard was the first to fly a liquid-fueled rocket, to build a multistage launch vehicle, to incorporate gyroscopes for stability in flight, and to add a nozzle to his combustion chambers to increase thrust. He was a tireless innovator, and in 1960 the U.S. government paid over to his widow and the Guggenheim Foundation a settlement of $1 million for infringement on Goddard's patents in the nation's development of military and civilian rocketry in the years after the Second World War. Newspapers that once might have referred to "moon-mad Goddard" now declared that the feds were paying off a debt related to "one of the costliest blunders in American history—disregarding concrete, patented plans to start building rockets . . . before the start of World War I."

Goddard saw what few others could, and in pursuing his visions, he both pioneered a new science and invented a field of engineering. He was a spark that others used to kindle a fire. Remembering the night they met, Lindbergh put it beautifully: "Sitting in his home in Worcester, Massachusetts, in 1929, I listened to Robert Goddard outline his ideas for the future development of rockets—what might be practically expected, what might be eventually achieved. Thirty years later, watching a giant rocket rise above the Air Force test base at Cape Canaveral, I wondered whether he was dreaming then or I was dreaming now."

3
Rocketry and Death

Three brilliant individuals from very different backgrounds provided ideas
that Nazi Germany turned into a terrifying weapon—the V-2 missile,
grandparent of all current space launch vehicles.

In those days, Robert Goddard may have been the only person in Massachusetts—
and quite possibly Vermont, for that matter—interested in the revolutionary
potential of rocket propulsion. But the world is a big place, and there were
minds in other corners of the globe thinking similar thoughts at around the
same time.

In Russia, the reclusive Konstantin Tsiolkovsky (1857–1935) imagined a future
studded with space stations and space colonies. He theorized that exploration
of the cosmos would improve the human race and that liberation from gravity would free people also from the shackles of social and economic oppres-
sion. The sooner we could get to space, then, the better. "Earth is the cradle of
humanity," he wrote. "But one cannot live in a cradle forever." Like Goddard,
Tsiolkovsky was a voracious reader, lonely as a child and largely self-taught.
He lost his hearing due to illness in his youth, and this further isolated him
from his peers. In his later years he sported spectacles and a salt-and-pepper
beard with streaks of white hair that curved like the sky in a Van Gogh land-
scape. Aside from teaching, writing, and dreaming about space, he spent years
fabricating metal dirigibles. Indeed, a contemporary photograph shows him
standing between his dirigible prototypes like a nineteenth-century beach
boy surrounded by surfboards. Most importantly, Tsiolkovsky derived the
formula we know today as the rocket equation, a statement of mathematical
tough love that allows for the calculation of the amounts of energy needed
to propel rockets of varying mass. Realizing the importance of the formula,
the Russian wrote down the date of its derivation: 10 May 1897. An oddity at
the time, a solution without a problem, the equation is now second nature to
any student of space exploration.

Tsiolkovsky was like Goddard in another respect as well. His ideas were so advanced that they were slow to catch on. But once people began paying attention, the significance of the solitary Russian's work became clear. In his homeland he is called the father of human space travel. In 2015 the Russian government named (or, more properly, *renamed*) a town in his honor. The largest crater on the far side of the moon is called Tsiolkovsky, as is an asteroid, 1590 Tsiolkovskaja. The man was a truly astonishing thinker.

Another rocketry pioneer was the Romanian-born German Hermann Oberth (1894–1989). Oberth was fascinated by the writings of Jules Verne and inspired by them to start designing vehicles that might one day be able to transport human beings to other worlds, as reflected in his 1923 work *The Rocket into Planetary Space*. Like both Goddard and Tsiolkovsky, Oberth seems to have conceived of the basis for cosmic transit almost entirely on his own. And his imagination was especially fecund. Despite the judgment of his superiors in the military and later in academia that his ideas were foolish and impractical, he dreamed not only of spaceships but also of space stations, space observatories, and even a cosmic doomsday weapon—a giant orbital mirror that could be tilted in such a way as to focus immense amounts of solar energy on a terrestrial target and thus burn it to bits. The obvious parallel is to the legendary solar heat rays invented by Archimedes and used to defend Syracuse against Roman incursions. But Archimedes needed a whole phalanx of small mirrors aligned to focus energy on a target. Oberth's plan required just one (although admittedly it was a very large one).

Oberth's career illustrates the point that much of the future of rocketry was theorized early on—in many cases before rockets were capable of doing much other than climbing fast and then exploding. For example, Oberth hit upon the idea of using planetary gravity to assist in accelerating spacecraft in transit, a concept now known as the Oberth effect. Another German, Walter Hohmann, conceived in 1925 of the "Hohmann transfer" as the most efficient means of using energy used in maintaining orbit around one celestial body to enter into the orbit of another. NASA flight planners still use these ideas when studying launch windows for various missions.

But scratching formulas on a chalkboard is one thing. Creating and harnessing the forces needed for space travel turned out to be a bigger engineering problem than a theoretical one. The members of the Baltimore Gun Club in Jules Verne's *From the Earth to the Moon* were missing limbs and jaws for a reason. Like artillery shells, rockets ride explosions, violent and convulsive chemical reactions that create intense heat and massive amounts of force. Rockets are value-pack volcanoes. Rockets shake the earth. Majestic as they appear from the outside, sentinels clad in white raiment, inside they're greasy metal tubes full of flammable toxic fuels desperate to escape their tanks. Suppressing this chemical insanity is an intricate labyrinth of fuel lines, turbopumps, valves, gauges, sensors, circuits, and switches straining to keep the liquids confined until the moment comes to slam the fuel and oxidizer together and ignite this pressurized witch's brew, creating a barely controlled catastrophe.

Humanity has always had big ideas. Capturing the flow of time in a system of symbols. Building the pyramids. Learning to fly. The big idea here is traveling to the stars. And what we've created to help us do it is a sort of Frankenstein's monster, big and dangerous and extremely temperamental. Rockets have a rap sheet. Rockets are killers.

The Funnel

Hermann Oberth was a soft-spoken man whose mustache and jowls drooped as if to illustrate the workings of gravity. He had the sleepy eyes of a born bureaucrat, but his brain rarely rested. Despite the fact that he never earned a doctorate, an important qualification in credentials-crazy Europe, he became a mentor to a number of German rocket-power enthusiasts. Some, like a dapper young Prussian aristocrat named Wernher von Braun, were members of the Verein für Raumschiffahrt, or spaceflight society. These dabblers traded ideas and theories about solar system exploration and built diminutive space darts in abandoned buildings on the outskirts of Berlin. Others, through the auspices of Opel-RAK, the Opel car company's futuristic propulsion project, designed and produced rocket-powered automobiles, trains, and at least one airplane in the late 1920s. Such vehicles were periodically rolled out and demonstrated to a fascinated public. For example, in June of 1928, the RAK-3 "rocket train" made its debut. The vehicle was really a modified train carriage with a car-like cockpit, outfitted with ten solid-fuel boosters—basically, large Roman candles—pointed rearward. The train hit a top speed of 160 miles per hour over a three-mile stretch of track, smashing the existing rail speed record as twenty thousand enthusiastic spectators looked on. On its second run, the vehicle jumped the rails and was destroyed.

It was Wernher von Braun, an accomplished theorist but also a gifted engineer, who was instrumental in creating the first fully functional, and thus reliably lethal, rockets. Like Goddard, he was captivated at an early age by the prospects of space travel. Unlike Goddard, though, he was a charismatic, socially adept individual who eventually had the financial resources of an entire nation to fund his work. He is perhaps best seen as a funnel, taking the various ideas of Tsiolkovsky, Oberth, Goddard, and others and producing a vile and beautiful machine as a result. Adolf Hitler's National Socialist German Workers' Party, the Nazis, consolidated their political power in Germany in 1934, as von Braun was finishing his PhD in physics at the Friedrich-Wilhelm University of Berlin. The German army offered von Braun financial support and physical facilities to continue doing professionally what he'd started as a hobby. Brushing aside any misgivings he might have had regarding the source of his funds, the young man eagerly accepted. In the years that followed, he and an ever growing team of engineers, mechanics, and managers turned the toys of gifted amateurs into serious weapons.

While von Braun and his friends in the spaceflight society would gleefully chase their early creations in cars and, on one occasion, caught a falling rocket by hand, the missile he ultimately produced for the Third Reich was no plaything. It stood forty-six feet tall and weighed fourteen tons when fully fueled. Von Braun called this monster the A-4. It ran on a mixture of alcohol and liquid oxygen, which was force-fed into its combustion chamber by two steam-powered rotary pumps that were themselves fueled by hydrogen peroxide. Perched above this propulsive machinery was the device's reason for being: almost two thousand pounds of TNT and ammonium nitrate, enough explosive material to shatter large buildings. As the A-4 progressed from a vision to reality, with many false starts and fiery crashes along the way, von Braun's reputation and status increased apace. He joined the Nazi Party. He became an officer in the German Schutzstaffel, or SS, the nation's homicidal elite force, and briefed Hitler personally on the merits of the secret weapon that German propagandists renamed the V-2. V-2 was the Third Reich's shorthand for Vergeltungsrocket 2, with *Vergeltung* meaning "vengeance" or "retaliation."

Hitler started the Second World War in September of 1939, when Germany invaded Poland on the basis of a phony border incident. While German forces were successful, smashing their way through both western and eastern Europe, there was little need for far-fetched military machines. Von Braun and his team were left to work in relative seclusion at Peenemünde, a secret facility on German's northern coast. But as the tides of war turned against the country, Hitler's interest in rocketry increased, eventually becoming a sort of mania. The Führer saw the v-2 as just the sort of secret weapon he needed to rally his followers and strike fear into the hearts of his many enemies. By 1943 he was urging the development and rapid production of missiles for strikes against the Allies. To speed matters up and to protect the project from Allied bombers, the Nazis built a large assembly facility called Mittelwerk, or Central Works, in an abandoned gypsum mine under the Kohnstein, a mountain in eastern Germany. The ss supplied labor—slave labor, as much of the work that went into building von Braun's war machines was eventually done by prisoners of the Third Reich, both foreign and domestic. Conditions in the Dora-Mittelbau camp that housed the missile facility's workers were unspeakably filthy and harsh. Disease and hunger ran rampant in the underground factory, and prisoners suspected of espionage were beaten, tortured, and often summarily hanged.

It might seem strange that the German military would characterize its airborne nightmares as weapons of revenge, given that the Nazis started the Second World War in the first place. But by the time the v-2 appellation was adopted, Germany was being systematically incinerated by Allied bombing raids. The name also played into German resentments over their treatment after World War I—though again, that was a conflict they were largely responsible for initiating. German aggression in the twentieth century was known not for the rigor of its logic but for the vehemence of its hatred.

Von Braun and his associates introduced a new species of horror to human warfare. The v-2 was the world's first long-range guided ballistic missile. Fired from a mobile platform, it could travel at speeds up to three thousand miles per hour for two hundred miles before crashing and detonating its explosive payload. It traveled so fast that there was effectively no defense against the new threat. All told, Germany launched three thousand v-2s, killing some nine thousand people in Great Britain and Belgium during the last year of the war. There was no tactical justification for these strikes. The missile wasn't accurate or powerful enough to incapacitate any significant military targets. It was, plain and simple, a weapon of terror, and it was employed to destroy and demoralize civilians. A v-2 strike in London in November of 1944, for example, destroyed a Deptford Woolworth store, riddled a city bus with shrapnel, and killed 160 people, including women and children. So much for the glory of the Third Reich.

In one of the most ambitious and difficult novels of the twentieth century, *Gravity's Rainbow*, author Thomas Pynchon explains that the innovative horror of the V-2 came from the fact that bystanders heard the missile explode on impact before hearing the sound of its approach. This disjunction of ordinary experience serves as a trigger for a central character's increasingly disordered mental state. And the rainbow—*gravity's* rainbow, the trajectory of the V-2 as it rises up from the continent, hits its apogee, and bends back down toward Earth on its way to burn and kill—arcs like a leaden parody of God's promise to man as represented in the Bible.

Guiding the Beast

Powering the V-2 was one challenge. Guiding it was quite another—and equally important. Von Braun and his team made significant progress in figuring out how to stabilize and steer their machines. Rockets could always be aimed. This requires little more than pointing. In the case of the early V-2s, this pointing was done by using a compass to figure the course from the launch site to the target. German rocket engineers would then calculate the amount of fuel needed to get the rocket to the apogee, or high point, of its trajectory, where the fuel supply would either run out or be cut off and the missile would start to return to Earth. Without some sort of ability to control the rocket, though, even small deviations from its course—caused by the wind, for example—would eventually lead to a large change in direction. The key to addressing these issues was the gyroscope, a stunningly simple instrument that reflects a natural law almost as important to space travel as gravity.

The gyroscope operates on the principle of conservation of angular momentum. Practically speaking, it means that once started, the spinning motion of a rotor around an axis will continue in the same path, and at the same angle, unless acted on by an outside force. If one suspends the spinning rotor in such a way that it can move independently of its housing, deviations in the relative orientation of the rotor and the housing can be measured and used to correct the path of the vehicle—here, the rocket—in which the housing is mounted. In the case of the V-2, electronic readings of such deviations were used to control graphite vanes in the rocket's jet stream and rudders attached to the four fins, which moved to keep the projectile on a more or less steady flight path. While the systems weren't perfect, and the V-2 wasn't terribly accurate, gyroscopes are what allow missiles to be "guided" at all. More sophisticated gyroscopes and sensors are used to this day in the guidance and orientation of rockets, airplanes, and even the International Space Station.

The Great Nazi Round-Up of '45

While the v-2 had little effect on the outcome of the Second World War, the rocket was nevertheless important as a demonstration of possibilities. This became crystal clear when a v-2 launched on a test flight in June of 1944 soared to a height of some 176 kilometers, or 109 miles. This was well into space by anyone's definition, and far beyond the capabilities of any other nation. The war ended less than a year later. Unsurprisingly, as conquering Soviet forces closed in on Germany from the east and Allied armies advanced from the southwest, intelligence officials in both the United States and the Soviet Union focused on securing as much of the dark magic of the Nazi rocket program as they could.

The secret American effort to recruit Nazi scientists and technicians was known as Operation Paperclip. It resulted in the postwar employment and repatriation to America of some 1,600 German scientists, engineers, and technicians, many of whom were Nazi Party members. The Soviets mounted a similar operation. A number of talented, Russian-born theorists and engineers studied rocketry in the USSR in the 1930s. They might well have produced a weapon similar to the v-2, except for the fact that Josef Stalin's paranoid regime turned against its best propulsion pioneers. The most talented of these engineers, Sergei Korolev, was imprisoned, tortured, and sent off to work in a gold mine. There he suffered a heart attack, almost starved, and lost all his teeth before being "rehabilitated" and eventually released. He went on to become the shadowy head of the Soviet space program—a man whose very existence was a closely guarded secret. In his book *The Right Stuff,* Tom Wolfe calls Korolev the Chief Designer, which is about all that anyone back in the fifties and sixties knew about the man.

The v-2 was an object lesson for Moscow. Though they missed out on the big fish, von Braun, the Russians "recruited" over 2,500 German scientific and technical personnel to assist them with their own rocketry program and other military research after the war. In 1946 many of these individuals were forcibly deported to the Soviet Union, where their work could be more closely monitored and controlled by Moscow. A number of German specialists worked to develop ballistic missiles for Moscow on an isolated island in Lake Seliger, half a day's drive northwest of the city. While the extent of German efforts in the USSR have not been well documented, it seems likely that German know-how contributed significantly to development of the Soviet Union's first intercontinental ballistic missiles. Indeed, according to von Braun biographer Michael Neufeld, the first project of the postwar Soviet missile program was to reconstruct and study von Braun's V-2s.

Von Braun was intensely aware of the value of his work. He distrusted his own government, which he knew was willing to execute him rather than let him divulge his secrets to an invading power. But he also feared the Soviets, whom he feared would treat him poorly as a prisoner of war—if he was allowed to live at all. Rather than take his chances with either the remnants of the Reich or the soldiers of the vengeful Red Army, von Braun chose to surrender himself and members of his engineering team to the U.S. Army.

It wasn't easy. The team hid its most important documents in a subterranean vault and then dynamited the entrance, burying it in rubble. They used subterfuge and forged documents to travel south, searching for Allied forces along the way. They finally managed to give themselves up on 2 May 1945, two days after Hitler killed himself in his Berlin headquarters and just a few days before the war officially ended.

The grizzled GIs who received von Braun's surrender were skeptical of the urbane, self-assured German's claims. He was Hitler's leading rocket scientist, he stated. He was the man who invented the V-2. "If we hadn't caught the biggest scientist in the Third Reich," one dogface recalled thinking, "we had certainly caught the biggest liar!" No one knew it at the time, but 2 May would prove to be an important date—crucial, in fact, to the development of the American space program. The previous night's sky had been completely black. This evening a sliver of moon appeared over the smoking ruins of Europe. It didn't provide much light. But for a handful of dreamers, it was a suggestion at least that renewal was real, that progress was possible, and that maybe— just maybe—humankind's future was to be found somewhere in the spangled sky, high above the butchered bodies and burned-out tanks.

But because these dreams were human, they also involved a practical consideration: Who would get there first?

4

The ss Major and the Suicide Squad

The end of the Second World War turned out to be the start of
another conflict, the Cold War—and the justification for an unusual
American experiment in situational ethics.

Understanding the development of the American space program in the latter half of the twentieth century involves reviewing the political conflict that
drove it. For centuries wars were fought over resources, religion, and the dynas-
tic pride of various broods of inbred dandies. The new source of conflict was
different. It was the Age of Ideology, and the world was riven by a simmering, occasionally deadly dispute: communism vs. capitalism, Eastern bloc vs.
Western alliance, the Soviet Union vs. the United States. Broadly speaking,
nations that supported the power of the state to create equality, security, and
some minimal standard of living for their citizens squared off against consti-
tutional republics, which insisted that individual rights—at least for certain
groups of individuals—trumped the prerogatives of a central government.
Neither side of the conflict lived up to its own ideals, but each was passion-
ate in pointing out the flaws of the other.

There have been other periods of history when the fate of civilization seemed
to hang in the balance. Europeans felt this way in the face of Muslim expansion in the Middle Ages. Persians spoke in apocalyptic terms during the Mongol invasions of their land in the thirteenth century. The aggressive regimes
of both Napoleon and Hitler inspired existential dread in their neighbors.
But there has never been a more multifaceted and wide-ranging struggle for
world dominance than the so-called Cold War between the United States
and its allies on one side and the Soviet Union and its satellite states on the
other. It wasn't just about each side's nuclear firepower, which increased during this period from merely nightmarish to a genuine threat to the survival
of humankind. John F. Kennedy characterized the conflict as "a race for mas-
tery of the sky and the rain, the oceans and the tides, the far side of space and
the inside of men's minds."

The Cold War began shortly after the end of the Second World War, when the United States, Great Britain, France, and the Soviet Union divided the conquered Germany into spheres of influence. While there was no single precipitating event, tensions rose as the Soviets consolidated their hold on East Germany in 1946 and made clear they weren't going to be leaving any time soon. Using intimidation, assassination, and espionage, they meanwhile installed puppet governments in several nations under their control: Poland, Czechoslovakia, and Hungary, to name a few. Soviet premier Josef Stalin predicted that war between the East and the West was inevitable. Addressing an audience gathered at Fulton College in Missouri in March of 1946, former British prime minister Winston Churchill spoke of an "iron curtain" being installed by the Soviets to divide Europe. The phrase caught on.

The United States was aided in the contest by years of intellectual emigration, voluntary and otherwise, from Europe. In the 1930s brilliant theoreticians like Albert Einstein, Edward Teller, Hans Bethe, and Enrico Fermi left the continent due in large part to anti-Jewish sentiment directed at them or members of their family. After the Second World War, America's covert induction of a whole host of German engineers and technicians added an instant network of rocketry experts. The assimilation of these disparate groups provided the country with the equivalent of what in these computer-crowded days would be called a massive processing upgrade. Combined with the nation's vast material resources and industrial know-how, the United States was well equipped for the coming competition.

The conflict lasted through 1991, when the Soviet Union disintegrated into a number of independent nations, including Russia, Belarus, Ukraine, and Georgia. Generally speaking, the Cold War roiled rather than raged. It was static. It was a vague fear, a nagging unease. But it was always tangible. To be labeled a capitalist in a communist nation like the USSR, East Germany, or the People's Republic of China during these years was a dire indictment. It was often a prologue to imprisonment, torture, or worse. To be a communist in America was a risk to one's employment status and prospects for social advancement rather than an invitation to bodily harm, unless one happened to live in Mississippi, but professing Marxism in any of a number of pro-Western countries like Angola, El Salvador, or the Philippines could readily result in a bullet to the brain. Soviet scientists invented missiles capable of flattening American cities. American engineers designed airplanes that could bomb

Moscow. Proxy wars raged in Africa, Asia, and the Middle East, on basketball courts and chess boards, at the Olympics and in the United Nations. It was a spectacularly complicated and sustained period of paranoia and dread. It was, many thought, the most important struggle humanity would ever wage.

Among the numerous contests to see which country—which "system"—was smarter and stronger than the other, the space race emerged as one of the most significant. There were military dimensions to the race, of course. Many in the Pentagon believed that the next war would be won by the nation that controlled the heavens. But there were other dimensions as well. Looking like the more progressive and modern-thinking society was a vital aspect of the global war for the hearts and minds of the developing world. In 1960 alone eighteen new nations came into existence, including Nigeria, Senegal, and the Democratic Republic of the Congo. The citizens of these countries were tasting independence from colonial overlords for the first time. They also controlled vast resources—oil, rubber, copper, and cobalt—desired by both the East and the West. The space race wasn't just about space. It was also about who could best influence populations right here on Earth.

The Proteus of Peenemünde

Enter—or perhaps, *reenter*—Wernher von Braun, one of the great case studies in the long-standing philosophical debate about whether important ends can justify repugnant means. Almost before the German guns ceased firing in May of 1945, U.S. intelligence operatives, acting on prompts from von Braun and his colleagues, were racing to recover all the information and hardware they could find at the Mittelwerk v-2 assembly facility. The Russians were scheduled to take control of the area in which the facility was located, so time was of the essence. Unsure of exactly what was important and what wasn't, the Americans took a hundred of everything they could find, loaded it up, and trucked it south, out of the grasp of the communists. Meanwhile, the German rocket engineers sat for hours of interrogation by American and British officers and officials. As it gradually became apparent that their stories were true, Washington began to appreciate that it was in possession of a remarkable windfall. Under the auspices of Operation Paperclip, von Braun and a number of his rocket-making colleagues were brought to the United States to assist in development of America's comparatively primitive rocketry program. In 1946 they were relocated to Fort Bliss, just north of El Paso in far west Texas.

It would be hard to imagine a more alien world than the Chihuahuan Desert for a group of northern Europeans like the v-2 engineers. It was essentially an open-air prison. Peenemünde, which served as headquarters for the Nazi rocketry effort during much of the Second World War, is located on the northern coast of Germany, where the rain and fog of the Baltic Sea support coniferous forests along the low-lying coastland. El Paso, one time Six-Shooter Capital of the West, is mountainous, dry, and hot—*very* hot. Cultural and social opportunities at the time were limited. It was the edge of the world.

"There's nothing in Texas," von Braun told an interviewer in 1959, thereby slighting one of the world's great assemblages of spiny plants, flesh-eating birds, and venomous reptiles.

"No," said the interviewer. "Of course not."

Valued for their expertise but distrusted for their affiliation with an enemy regime, it was occasionally unclear whether the Germans were dangerous detainees or distinguished guests. Von Braun characterized himself and his colleagues as "prisoners of peace." Here and at the nearby White Sands Proving Ground, in southeastern New Mexico, they instructed U.S. Army personnel in German rocketry techniques. As the Cold War began to heat up, the Germans became increasingly valuable to the American military. The Pentagon was reluctant to advertise its reliance on the Germans, of course, and for good reason. For the four years from 1941 to 1945, the Germans had been brutal and intimidating military adversaries. Worse, and as became increasingly apparent, they were perpetrators of unspeakable horrors against huge numbers of defenseless noncombatants. The case against von Braun in particular is instructive in this regard. First, he'd built weapons that were used to kill British and Belgian civilians. This was an obvious charge, and easily proven. But the Germans had no monopoly on the slaughter of noncombatants during the Second World War. The United States killed men, women, and children as well, most notably in Dresden, Hiroshima, and Nagasaki. Von Braun met such charges head on by stating that it was wartime and he was defending his country, glossing over the inconvenient fact that Germany wasn't a victim of the war but rather its prime mover.

Aside from the damage caused by the German military's use of the v-2, however, was a second set of horrors: von Braun and his rocket men benefited from the assembly work of prisoners who were subjected to appalling conditions that only deteriorated as the war ground on. These prisoners—Russians and Poles, Germans and French nationals, some but not all of them Jewish—suffered from exposure, starvation, disease, brutal beatings, and summary execution. Indeed, it's sometimes said that more men died building the v-2 than were killed by its impacts. Von Braun's defense here was that he had no hand in the atrocities and that his knowledge of these work conditions was limited. This question has never really been settled. There is no evidence that von Braun himself tortured or killed anyone. On the other hand, it's hard to believe that someone of his intelligence and access could really have been blind to the brutal methods employed to assemble his precious death machines.

As the extent of Nazi atrocities before and during the Second World War became clear during the 1950s, Western sentiment hardened. It was difficult for many Americans to understand why the nation should welcome as an ally a man who was instrumental in the development of such a fearsome and indiscriminate weapon. Tom Lehrer's famous 1965 song about the man summed up the matter well, though perhaps unwittingly stating von Braun's case for him—that he wasn't hypocritical, just *apolitical*.

For his part, von Braun claimed that his participation in the Nazi war effort was mandatory, not elective. To disobey would have meant imprisonment or possibly death—and indeed, he was briefly detained by German authorities in March of 1944, accused of the vague crime of "defeatism" for wanting to build a rocket that could reach the moon, rather than Mons. The extent of his culpability for the crimes of his government has for years been vigorously debated. What is not debatable is the importance of his work, and the work of his colleagues, in getting America into space. He came complete with not only practical experience but also a head full of huge ideas: multistage "piloted rockets," space stations and jet packs and—naturally—Hermann Oberth's Giant Orbital Mirror of Death. Tactically, the decision to incorporate the former s s officer into the American rocketry effort rather than imprison him or leave him to the Soviets paid off. The German's Jupiter-C launch vehicle, the grandchild of the v-2, lifted America's first satellite into orbit. More importantly, von Braun was also the head honcho behind the Big Boss—the Saturn V rocket, the storied heavy-lift, multistage sky slasher that sent astronauts Neil Armstrong and Buzz Aldrin on their way to the moon in 1969.

The truth is that the prospect of Moscow beating Washington in the contest to claim the heavens seemed more important at the time than holding von Braun accountable for his service to a criminal regime. Justifiably or not, he was embraced by the U.S. government. And he hugged it right back. As big as John Wayne, movie-star handsome and as sleek as an otter, von Braun became an enthusiastic American and a born-again Christian, a famously debonair emissary from the future.

Von Braun's metamorphosis from little-known wartime refugee to America's foremost space celebrity in the 1950s was nothing short of astonishing. Over the course of twelve years after he arrived in secret in the United States, he took over day-to-day management of the army's ballistic missile program, wrote influential magazine articles, partnered with Walt Disney on television to present his plans for America in space, and built the rocket that put the first American satellite in orbit. He did so with personal charm, organizational genius, and a firm but entrepreneurial management style that inspired loyalty and collaboration in his American and German employees alike. He was admired and widely respected, even if elements of the population never quite trusted him. Of course he was debonair and well spoken. *All Nazi officials were debonair and well-spoken!* This weird ambivalence surfaced in 2023 in the fifth Indiana Jones movie, *Indiana Jones and the Dial of Destiny*, in which a thinly disguised von Braun–like character, having dutifully placed American astronauts on the moon, schemes to turn back the hands of time so that Nazis like him can prosecute their conquest of the world more successfully.

Von Braun wasn't the only German with a questionable past welcomed by Washington. He was only the most protean—and the most successful. Hermann Oberth, sleepy-eyed recipient of the Third Reich's gratitude for his military work during the Second World War, wound up in America too, helping his former acolyte von Braun in his rocketry development. Kurt Debus, who became the first director of launch operations at the Kennedy Space Center, was also a former Nazi and ss officer who traded his death's-head insignia for a fridge full of Bud and a pair of bowling shoes. Arthur Rudolph helmed the Saturn V development project at Marshall Space Center, winning the NASA Exceptional Service Medal before eventually—well after his services were needed, that is—being investigated for war crimes by the U.S. government.

Perhaps the most notable of the tainted Germans after von Braun was Dr. Hubertus Strughold, who came to America under the auspices of Operation Paperclip in 1947. Strughold worked for many years at Brooks Air Force Base in San Antonio, where he earned professional fame for his contributions to the space program, including input on the design of pressure suits for the astronauts, and became known as the Father of Aerospace Medicine. He is also said to have coined the term *astrobiology* and created the first "Mars jars," low-tech simulators of the Martian environment used by scientists to study the possibilities for life on the Red Planet. Despite his accomplishments, Strughold never quite managed to outrun his past. Over the years, he was alleged to have associated with German officials who conducted horrific experiments on human subjects taken from the Dachau concentration camp. A later accusation that he orchestrated or at least countenanced one such experiment led to a substantial revision of Strughold's legacy—including discontinuation in 2013 of the Space Medicine Association's annual Hubertus Strughold Award for achievement in aerospace-related medical research, an honor that was bestowed several times on American astronauts. The allegations against Strughold were never proven in court. The fire was hard to see. But the smoke was thick enough that ignoring it eventually became impossible.

Never Say "Rockets"

As important as German know-how proved to be, the United States wasn't starting from scratch in its postwar rocketry work. Incorporating the studies of Frank Malina and his team of rocketry buffs associated with the California Institute of Technology, the army had built and by 1946 was testing the WAC (without attitude control) Corporal, a small but reliable liquid-fueled launch vehicle.

Malina was born in Texas to Czech immigrants. He attended Texas A & M University, where he studied engineering and played in the Fightin' Texas Aggie Marching Band, which doesn't actually do much fightin' but nevertheless wears military-style uniforms and is famously large and loud. Malina, trim and soft spoken, a romantic at heart, never quite fit in. After graduation, he headed west in 1934 to study airplane design at Caltech. There he blossomed under the tutelage of Professor Theodore von Kármán, namesake of the Kármán line and often cited as one of the most brilliant thinkers of the age. Certainly von Kármán himself thought so; he once ranked himself just behind Sir Isaac Newton and Albert Einstein for the quality and originality of his work. When Malina became interested in rocketry, still considered a dubious area of study for serious scientists, von Kármán helped as best he could, offering the young grad student the use of on-campus facilities to conduct his experiments. At around this time Malina met and began working with a machinist, Ed Forman, who became the group's mechanic and metal worker, and an amateur chemist, Jack Parsons, who specialized in blowing things up. Parsons was made the group's rocket-fuel expert. He had other interests as well. A large, good-looking man with a head of jet-black hair that rose like a storm cloud above his head, Parsons fancied himself a sort of latter-day Lord Byron—mad, bad, and dangerous. He drank, experimented with drugs, and dabbled and later dived headlong into an odd and occult theology not easily summarized in the pages of a family-oriented rocketry primer. At least one friend—Forman—thought that Parsons had opened a door to the underworld as a result of one of his arcane rituals. For years afterward, Forman shuddered when he recalled the incident.

Members of the group, including Malina's fellow student, mathematician Tsien Hsue-Shen (sometimes spelled Qian Xuesen), were inspired by Robert Goddard's ideas but unable to collaborate with him, as Goddard was wary of sharing details of his work. Undeterred, the men began cobbling together their own rockets in the 1930s in a dusty arroyo outside of Pasadena. The reliability and accuracy of these early devices is reflected in the name bestowed by Caltech students on Malina and his friends. They were known as the Suicide Squad. But the rockets got better. As the Second World War heated up, the Army Air Force contracted with the squad for production of solid-fuel rockets called JATOs (jet-assisted takeoff rockets) that could be mounted to aircraft and fired to assist the planes in taking off quickly. Parsons developed a solid-fuel rocket that would burn evenly and consistently, using asphalt in his JATOs to pave the way, so to speak, for a successful and ultimately quite lucrative relationship with the Pentagon. The JATOs led to experiments with liquid-fueled motors and eventually from aircraft into rockets that aimed considerably higher than contemporary airplanes could reach. Malina and his squad were instrumental in creating a series of sounding rockets called the Private, which were tested in California and, later, west Texas. A version of their WAC Corporal, next in the evolutionary line, was the first U.S. rocket to fly higher than fifty miles and thus reach space.

Despite his successful career in rocketry and contributions to the U.S. war effort during the Second World War, Malina was investigated during the postwar Red Scare years by the FBI, which suspected that he was a member of a local cell of the Communist Party. Though he had in fact espoused left-wing sympathies at various times in his career, there was no evidence that he'd done anything other than exercise his First Amendment rights in doing so. Nonetheless, he and Tsien, who had been born in China, were harassed by the federal government in various ways for their alleged political affiliations.

Each man ended up leaving the United States. Malina moved to Europe, where he worked for a time for UNESCO and eventually became an artist of some renown. Tsien was at one point detained by federal authorities, who believed him to be a communist spy. He denied such allegations and was backed up in his denials by fellow academics, including von Kármán. Some observers pointed out that Tsien had worked to develop technology to aid U.S. combat aircraft during the Second World War. He had also held the temporary rank of colonel in the U.S. Army when he journeyed to postwar Europe to interview German rocket scientists, including Wernher von Braun, for possible admission to the States. This record of service seemed to make no difference. After years in legal limbo in the United States, and active FBI surveillance, Tsien was allowed to "self-deport" himself to China in August of 1955, as part of a deal involving the release of American servicemen captured during the Korean War. Embittered by his experience with McCarthyism, Tsien went on to become an important contributor to the Chinese space program—the father of modern Chinese rocketry, as generations of Chinese students have been taught. He survived the venomous political backbiting of the Cultural Revolution, passed away in 2009, and is now a secular saint, with his accomplishments enshrined in a Shanghai museum featuring what one source claims are "70,000 artefacts." Scholars doubt that Tsien was really an agent of Beijing, which means that chasing him out the country and into the arms of Mao Zedong could be considered an embarrassing "own goal" by American security officials. The story remains a fascinating subplot to both the Red Scare days in the United States and the story of the American space program.

The Suicide Squad's work in rocketry was institutionalized in both Caltech's Jet Propulsion Laboratory, or JPL, an important component of NASA, and the Aerojet Engineering Corporation, which members of the squad formed to sell JATOs to the Army Air Corps. Note the use of the word *jet* in both names. In reality, neither organization built jets. Rather, the term was used because rocketry was still a fringe pursuit, and the organizers of both JPL and Aerojet didn't want to put off possible customers and investors by suggesting they were attempting to construct devices most Americans still thought best belonged in the funny papers. JPL went on to build America's first satellite,

THE SS MAJOR AND THE SUICIDE SQUAD

Explorer 1, and dozens of important and advanced robots and space probes since. Shares in Aerojet, which eventually merged with other companies and is now known at Aerojet Rocketdyne, eventually made Malina a wealthy man.

Self-taught rocket-fuel expert Jack Parsons also faced FBI investigation—though the reasons for this related as much to his cult activities as to his political affiliations. His life could fill the pages of a complicated novel, involving the names of not only Malina and von Kármán but also that of science fiction writer and Scientology founder L. Ron Hubbard, with whom Parsons became entangled after he was kicked out of JPL and Aerojet due to what one source refers to as his unorthodox beliefs and reckless work behavior. Parsons never stopped tinkering with rockets and explosives, however. It was during one such experiment in 1952 that he died as a result of an explosion. Exactly how that happened has never been completely explained, as perhaps is fitting given the complicated life of one of space exploration's oddest pioneers.

Into the Fray

Ironically, while talented rocketeers like Malina and Tsien were chased out of the country because of their socialist and pro-labor sentiments, men who had helped Adolf Hitler wage war on the United States and kill American allies were meanwhile enlisted to help the federal government build the nation's future in space. Using hundreds of V-2 parts seized near the end of the war and shipped to far west Texas by the U.S. Army, the German team set about resurrecting, and improving on, the V-2. Despite the challenges of starting over again in a new facility with these odds and ends, Von Braun and his cadre managed to construct and launch some sixty-four rockets between 1946 and 1952. In fact, the first photographs of Earth from space—sixty-five miles up, just above the Kármán line—were taken from a V-2 in October 1946. The team also managed to initiate the first American missile attack on Mexico, when they accidentally sent a V-2 arcing up over the border to bury itself in a mountainside outside of Juárez. Thankfully, no one was hurt.

Many of the Fort Bliss Germans and their families were eventually offered government employment contracts and moved to the army's Redstone Arsenal facility in Huntsville, Alabama. Here, working with American engineers, they developed the Redstone rocket, a lineal descendant of the V-2. The larger and more powerful Redstone was first test-fired in 1953. By 1958, after many further trials, a Redstone climbed to an altitude of forty-seven miles, at which point its nuclear payload was detonated, creating a fireball over the Pacific Ocean that could be seen from some eight hundred miles away. This was reckoned by some to be progress. But half a world away, the Soviets were making progress of their own—and they were moving faster than the United States. The world would soon take notice.

The All-Time Greatest
Space Exploration Playlist

1. David Bowie, "Space Oddity": One of the greatest of many pop songs about slowly going bonkers, this one's redeemed by Bowie's curious, almost triumphant ownership of his narrator's mental deterioration. Or evolution. In this case, it's hard to know which.

2. Lightnin' Hopkins, "Happy Blues for John Glenn": Houston's legendary guitar wizard isn't completely clear on the motivations for Glenn's Mercury-Atlas 6 flight, but no matter. The pride shines through in this paean to NASA's 1962 orbital achievement.

3. John F. Kennedy, 12 September 1962 Rice University speech: Absolutely required listening. The movie star president throws down the technological gauntlet in a Houston speech at the dawn of the decade. An energized America picks it up, puts it on, and starts welding.

4. Elton John, "Rocket Man": The beauty of this little gem derives not only from Sir Elton's orbital voice but also from its evocation of the psychological stresses felt by a seemingly average man who's just trying to do his job while wrestling with enormous external expectations. We can relate. There's a lot of science we don't understand as well. And it's true: Mars ain't the kind of place to raise your kids. Not yet, anyway.

5. Dave Giles, "The Last Man on the Moon": This sweet-natured slice of guitar pop is a heartfelt tribute to astronaut Gene Cernan, penned and performed by the British singer/songwriter who fell in love with the American space program at an early age and has never stopped believing.

6. Dua Lipa featuring Da Baby, "Levitating": In a video produced for the song, the Anglo-Albanian chanteuse solicits the attentions of

an unspecified sugar boo while embarking on what appears to be a space elevator, traveling upward to Roller World, where, amid much dancing and festivity, zero gravity is either achieved or briefly simulated. Okay, so it's not really about rockets and the like per se, but if you can listen to this catchy space-themed jam without seeing a few stars, you don't need to be reading about the *Voyager* probes. You need to be leaning into your accounting classes.

7. Public Service Broadcasting, "Go!": Just one of a bunch of Public Service Broadcasting tunes that could be included on this list, "Go!" samples mission control colloquies to bring us a driverless car of a tune that carries some real joy—and palpable call-and-response tension.

8. "Jupiter, the Bringer of Jollity": You can't really dance to the fourth movement of Gustav Holst's immensely popular orchestral suite *The Planets* unless you just don't care what people think of you anymore, and because Jupiter is the source of intense radioactivity, it wouldn't be bringing any visitor jollity for long. Nevertheless, this magisterial aural imagining of our biggest planet's supposed astrological characteristics—Holst was a horoscope fetishist—is so impossible to ignore that many contemporary composers have given up trying. One of its companion pieces, "Mars, the Bringer of War," actually shows up in more movies and television shows but is probably best reserved for moments when you don't mind feeling as if someone is chasing you with a battle axe.

9. Frank Sinatra, "Fly Me to the Moon": Ol' Blue Eyes would have lasted about ten minutes in an Apollo command module before he discovered there was neither ice nor ashtray on hand. Then he would have flipped mission control the bird and exited in search of the nearest hotel lounge and a bar full of broads. Still, the yearning, burning—and phrasing—involved in the notion of space travel as the ultimate expression of romance have never been better.

10. John Craigie, "Michael Collins": "Sometimes you get the fame, sometimes you sit backstage. But if it weren't for me, them boys would still be there." Indeed they would, though it's hard to imagine Collins, NASA's renaissance man, doing anything unseemly. This sim-

GREATEST SPACE EXPLORATION PLAYLIST | 41

ple, funny song gives *Apollo 11*'s unsung hero a well-deserved three minutes in the limelight.

11. The b-52s, "Planet Claire": About as campy as it's possible for anything not associated with the old *Batman* TV series to be, "Planet Claire" is danceable, Tang-flavored nonsense, deeply flawed astronomy for the fun of it. Who knew?

Also stellar and worth a play: Ash, "Girl from Mars"; The Tornados, "Telstar"; Earth, Wind & Fire, "Shining Star"; David Bowie, "Starman"; and The Church, "Under the Milky Way."

Best to avoid: Burl Ives, "The Tail of the Comet Kohoutek."

5

A Starting Gun Called Sputnik

The United States and the Soviet Union used space technology to fight
a public relations war in the heavens. For years it looked like the Soviets were winning.
Then an American president issued a challenge.

Testing means of mass destruction wasn't the only science being done in those days.

Indeed, despite its pop culture reputation as a buttoned-down, unimaginative era, dominated by The Man in the Gray Flannel Suit, the fifties were years of remarkable innovation and inquiry. Case in point: the 1958 International Geophysical Year, which, confusingly, wasn't a year at all but ran for a period of eighteen months, from 1 July 1957 through 31 December 1958. The IGY was established by scientists to rekindle the spirit of the International Polar Years of 1882–83 and 1932–33, which generated international exchanges of information regarding the planet's North and South Poles. The 1958 project had a broader focus, one that incorporated all manner of planetary investigations. It was meant to be the general session of a sort of United Nations for Earth researchers, an opportunity to swap data and theories and perhaps achieve consensus on outstanding issues in a number of fields.

In the spirit of the IGY's aims, President Dwight D. Eisenhower announced that the United States would launch a satellite sometime during the extended year. Once established in orbit, the satellite would transmit geodetic data back to Earth to share with scientists around the world. To the intense annoyance of Wernher von Braun and his U.S. Army team, who considered themselves the nation's premier rocketeers, Washington chose the navy's Vanguard launch vehicle over both the army's Redstone and a fledgling air force project to accomplish this important feat.

The IGY was a success. Sixty-seven nations, including the United States and the Soviet Union, participated. Among other achievements, a number of countries agreed to the Antarctic Treaty, which governs the use and habitation of the South Pole and is still in effect today. Joint British-American submarine reconnaissance mapped undersea ridges in the Atlantic Ocean that confirmed new thinking about tectonic activity.

As important as they were, though, such developments have always been overshadowed by a beach-ball-sized satellite launched by the Soviet Union on 4 October 1957. Called Sputnik, or "fellow traveler," it was the first manmade object to orbit the earth. Like Goddard's rocket thirty years earlier, it changed the world. Unlike Nell, though, Sputnik's effects were immediate and obvious. The Russians had supplanted the Germans and the Japanese as America's sociopolitical bogeymen soon after World War II. Watching their aggressive brand of authoritarian communism, the United States came to believe that it was locked in a cage match for control of the globe with the soulless fanatics of Moscow. Most Americans simply assumed that their country was bound to prevail. Sputnik, then, was a slap in the face, a highly visible and embarrassing rip in the nation's assumption of scientific preeminence. "RUSSIANS LAUNCH FIRST SPACE SATELLITE; CIRCLING EARTH AT 18,000 MILES AN HOUR" announced the *Washington Post*. "THE SPACE AGE IS HERE" trumpeted New York City's *Daily Express*. The *Cedar Rapids Gazette* led with the somewhat cryptic "SOVIET: FIRST STEP TO THE MOON."

Laymen lamented that the Soviets had launched a satellite before we did. Military and aerospace insiders were less impressed with the fact that Moscow had accomplished orbit first than with the size and weight of the Russian satellite, which, at almost two hundred pounds, dwarfed anything the United States was preparing to launch. The obvious takeaway was that the Soviet Union was far ahead of America in missile-building technology. With one swift stroke, the Red Menace had established a significant and alarming lead in the technologies of orbital mechanics, remote surveillance, and, presumably, mass destruction.

Sputnik wasn't an immediate threat. It didn't carry a weapon or even a payload, aside from its radio transmitter and the batteries needed to power it. In fact, the only damage any of Moscow's early satellites did to the United States occurred in 1962, when a twenty-pound piece of Sputnik 4 smashed into the middle of a street in Manitowoc, Wisconsin, a community known for the manufacture of both submarines and cheeses, presumably in separate facilities. No one was injured by the impact, which gouged a good-sized hole in the asphalt. The site is now a minor tourist attraction. But actual armaments weren't the point. It was the implications of Sputnik that mattered.

Maybe the communists really *were* going to bury us, as Soviet first secretary Nikita Khrushchev famously vowed in 1956.

Forward-thinking analysts had warned that the world might think ill of American technological capabilities if the nation wasn't a leader in space exploration. But it turned out the world wasn't needed. In the wake of Sputnik, the nation did an excellent job of castigating itself, embarking on an orgy of self-criticism in the newspapers, on television, and even in the halls of Congress. Our lives were too easy, we realized. Our children were soft-headed layabouts. We'd been caught drowsing, sedated by sitcoms, dreaming of shark-finned Cadillacs cruising endless interstates. Suddenly, the sky was occupied. There was a communist spy in America's attic. It was time to wake up. And just as a slumberer often does when roused from sleep, the country stumbled as it took its first steps.

Imagination Meets Reality

It wasn't that Sputnik's success was inconceivable.

American scientists, military men, and civilian policymakers were keenly aware that the Soviets might deploy the world's first satellite, and they had fretted for years about the possibility of socialist domination of space. The Rand Project, forerunner of the Rand Foundation, cautioned against the prospect of Soviet primacy in the heavens as early as 1946. In a report commissioned by President Harry Truman in 1952 (but not issued until 1953), nuclear chemist Aristid von Grosse opined that the engineering knowledge required for putting an unmanned object in orbit already existed, though more powerful rocket boosters would need to be developed. He noted that "a design for such a large stage was already on the drawing boards of Dr. von Braun and his associates in Peenemünde, Germany, in 1945. This German project ... was designed for transatlantic bombing of the United States." Dr. von Grosse listed several uses to which an unmanned satellite could be put and noted that if the Soviets were to be first into space, "it would be a serious blow to the scientific and engineering prestige of the United States the world over. It would be used by Soviet propaganda for all it is worth."

By 1953 the Soviets had developed their own atomic bomb—thanks in part to spying on U.S. weapons programs—and were just a few months away from successfully detonating a hydrogen weapon. Anxieties in Washington only increased when, in August of 1957, Moscow successfully tested the R-7, the world's first intercontinental ballistic missile (often known by its shorthand title, the ICBM). Simply put, an ICBM is a really big missile capable of leap-frogging the oceans that have traditionally kept North America safe from hostile foreign governments and upon arrival detonating a nuclear device capable of destroying whole cities. Analysts knew that a missile that could cross the Atlantic with a sizeable payload might easily be modified to create a rocket that could haul a satellite into space.

Unease about a spacefaring future was already percolating in popular culture. While the idea of competition with the Soviet Union, an American ally during the Second World War, was relatively new, the possibility of rocket-powered space travel had been a staple of newspaper headlines, pulp fiction, and Hollywood hits for years. Indeed, even before Robert Goddard fired his first liquid-fueled dart in 1926, German scientists and pilots were touring the United States and Canada, touting the possibilities for rocket travel to the moon. Later in the decade, Hermann Oberth proposed deliveries of mail from Germany to America by rocket. A U.S. diplomat said the project sounded promising but that Washington would have to study the safety of the plan before agreeing to let the Eastern seaboard become a postal pin cushion. "GERMAN AVIATION CIRCLES STIRRED BY ROCKET PLANS" read an unintentionally ominous 30 September 1929 headline in the *Montreal Gazette*. Reflecting such excitement, comic strip heroes Buck Rogers and Flash Gordon, science fiction analogs of Western cowboy heroes, were soon soaring through the heavens in spaceships and jet packs to combat intergalactic evil. The most popular toy of the 1934 Christmas shopping season, space historian Margaret Weitekamp writes, was the "Buck Rogers ray gun" ("LIMIT ONE PER CHILD. PHOTON EMISSION STIMULATOR SOLD SEPARATELY!").

The Second World War made the destructive capacity of rocket-borne weapons clear, most obviously with the V-2 but also in connection with Germany's plans, never realized, to use a radio-controlled rocket to attack the United States. The notion that a foreign power could someday use this method to deliver an *atomic* weapon, like the ones that had destroyed Hiroshima and Nagasaki in 1945, caused nightmares among both experts and laymen. After a rash of "flying saucer" sightings all across the United States in the summer of 1947, a small but insistent portion of the American population came to believe that extraterrestrial beings had visited Earth and might well do so again. Compared to such matters, a future featuring rockets wasn't much of a stretch. Then, too, in the late forties, Robert Heinlein, Isaac Asimov, and Arthur C. Clarke began to attract readers from the general public with their science-based tales of far-flung space civilizations and interstellar travel. Suddenly sci-fi—originally called "scientifiction"—was no longer a guilty pleasure, a haunted forest of tentacled eyeballs and helpless, half-dressed Earth maidens you didn't want your employer to know you enjoyed reading about. In the hands of the Big Three, sci-fi shunned schlock in favor of technologically sophisticated, though still plenty strange, speculations about what the day after tomorrow might bring. Even Wernher von Braun got in on the act. All during the Second World War, he'd carefully kept his subscription to the sci-fi magazine *Amazing Stories* active and had intermediaries forward it to him from a post office box in Sweden. Underutilized as a so-called prisoner of peace at Fort Bliss in far west Texas, he wrote a long novel in the late 1940s about a mission to Mars, which he augmented with a 120-page technical appendix intended to demonstrate the viability of such a trip. The whole thing seemed less like a novel than a blueprint.

Not all futurist musings were as well thought out. A 1949 Republic Studios serial, *King of the Rocket Men*, spawned a slew of sequels, including the immortal *Zombies of the Stratosphere*, in which the prosaically named Larry Martin uses his jet pack ("Down" and "Up," say the controls) to thwart a gang of Martian wise guys who want to shove their own planet into Earth's orbit, breaking all the laws of God and man. (Look closely and you'll see Leonard "Mr. Spock" Nimoy in an early role as a stratospheric zombie.)

But that was only one end of the cinematic spectrum. In the 1950 film *Destination Moon*, a serious attempt to depict a crewed rocket trip to the lunar surface and the reasons for the journey, a character known only as "General Thayer" describes the development of "satellite rockets" as "an absolute necessity" for the United States. "The race is on," he tells an audience of concerned American industrialists. "And we'd better win it, because there is absolutely no way to stop an attack from outer space. The first country that can use the Moon for the launching of missiles will control the Earth. That, gentlemen, is the most important military fact of this century."

In a more contemplative mood, Ray Bradbury's great collection of short stories, *The Martian Chronicles*, published in 1950, leads off with a tale called "Rocket Summer." In the story, the first crewed mission to Mars is preparing to leave Earth. Public enthusiasm reaches dangerous levels, but the president of the company that built the launch vehicle has serious reservations about entrusting this new technology to a world that hasn't yet learned to curb its destructive and wasteful tendencies. Humanity isn't ready to meet the universe, thinks the CEO as he looks out over the launch site. And yet veneration for the new technology grows louder.

> "Yezzir! Yezzir!" he heard the far-off, faint and raucous declarations of the vendors and hawkers. "Buy ya Rocket Toys! Buy ya Rocket Games! Rocket Pictures! Rocket soap! Rocket teethers for the tiny-tot! Rocket, Rocket, Rocket! Hey!"

Walt Disney, America's family-friendly fantasist, was riding both scientific and popular momentum when he partnered with von Braun in the midfifties for a series of television presentations explaining the principles and prospects of space exploration by rocket. Disneyland's Rocket to the Moon attraction, featuring a scale-model Trans World Airlines–sponsored space vessel, debuted in 1955. Signage at the base of the model explained that "the full-scale ship would be 240 feet in length and designed to use nuclear energy as fuel. Stabilized in flight by gyroscopes, it would be controlled by automatic pilots and magnetic tapes. Landing tail-first, no air-foils or wings would be necessary, its vertical descent controlled by its jets. The three retractable landing legs would be equipped with shock absorbers."

All of this is to say that the fantastical possibilities of space travel—and space combat—lurked in the American mind long before Sputnik flew. Nevertheless, the reality of Soviet success proved unsettling. This may have been in spite of the monsters-and-flying-saucer fantasies of popular culture. On the other hand, it may have been *because* of them. After all, if rocket ships could enable Flash Gordon's nemesis, Ming the Merciless, to rule the planet Mongo, who knew what the Kremlin might be able to accomplish with the technology. And where was Larry Martin when we needed him?

Kaputnik

Sputnik eventually fell back into Earth's atmosphere, burning up on its way toward the planet's surface. But the aftershocks lingered. As the National Security Council declared in June of 1958:

[The] facts which confront the United States in the immediate future are (1) the U.S.S.R. has surpassed the United States and the Free World in scientific and technological accomplishments in outer space, which have captured the imagination and admiration of the world; (2) the U.S.S.R., if it maintains its present superiority in the exploitation of outer space, will be able to use that superiority as a means of undermining the prestige and leadership of the United States; and (3) the U.S.S.R., if it should be the first to achieve a significantly superior military capability in outer space, could create an imbalance of power in favor of the Sino-Soviet Bloc and pose a direct military threat to U.S. security.

As jarring as the whole event was for the American psyche, there was a silver lining. By flying its cosmic beach ball over the continental United States, the Soviet Union won a short-term propaganda victory but set a precedent that would cost it dearly in the long-term surveillance wars it would go on to fight with the United States. The precedent was that outer space—as opposed to "air space," or the *skies*—was freely navigable, regardless of whose national boundaries one crossed. Thus, while the Russians could shoot down CIA pilot Francis Gary Powers as he crossed over the USSR in his U-2 spy plane at an altitude of seventy thousand feet in 1960, it had no basis in international law for attempting to damage, or even complain about, a U.S. satellite transiting Russia at 150 miles up, well above the Kármán line. As American satellite and photography technologies gradually improved over those of the Soviets, this "free space" precedent became increasingly important.

But the American public wasn't interested in the vagaries of international law. It had been punched in the gut. It wanted to punch back. It wanted a *response*. This visceral craving only intensified in November, when the Soviets sent up a second satellite, Sputnik 2, which was much larger than the first and carried a canine passenger named Laika, who died during the flight. The United States attempted to answer the challenge a month later. On 6 December 1957 the navy launched its Vanguard TV-3 rocket, which carried a grapefruit-sized artificial satellite similar in appearance to Sputnik, though much smaller. It was a short flight. The rocket rose just a few feet off the launchpad before losing power. It then fell back to Earth and exploded. The tiny satellite was thrown clear and seemed to flee into the surrounding swampland, emitting a plaintive radio signal that was clearly audible but difficult to track down. The fiery accident was televised later that day and proved to be almost as big an embarrassment for the country as being beaten by Sputnik in the first place. America's runaway satellite was dubbed Kaputnik in the press. At the United Nations, a Soviet diplomat inquired publicly whether the United States would like to avail itself of the financial aid his country made available to governments of undeveloped nations.

This wasn't exactly the response the public had been looking for.

Fortunately, the nation had a plan B. After the explosion of the Vanguard TV-3, Washington turned to Wernher von Braun's army team, which had doubted the success of the Vanguard project all along and was happy to announce that it just happened to have a rocket in readiness to replace the navy's wayward booster. So it was that the U.S. Army Ballistic Missile Agency launched America's first satellite, Explorer 1, on 31 January 1958. Constructed by the Jet Propulsion Laboratory, Explorer 1 was just under seven feet in length, weighed thirty-one pounds, and rode atop the army's Jupiter-C rocket (a modified Redstone) launched from Cape Canaveral.

The launch wasn't just a political response. Pursuant to Eisenhower's commitment to the International Geophysical Year, Explorer 1 carried a "cosmic ray detector" (a Geiger-Mueller radiation sensor) that provided data indicating the presence of radiation trapped in Earth's orbit by the planet's magnetic fields—what we now call Van Allen belts, after the scientist who identified them. The extended orbit of Explorer 1, the first of fifty-five Explorer satellites to be launched over the next seventeen years, also showed that micrometeoroid damage to satellites was less likely than some had feared—good news for any would-be spacewalkers wary of being torn apart by bits of orbital debris careening around the planet at thousands of miles per hour.

The Birth of NASA

Washington had for years been content to leave rocketry research to the armed forces—when it thought about the subject at all. No longer. Despite the timely success of Explorer 1, politicians of all stripes agreed that in light of Soviet successes, it was time to make the space race a national priority, under centralized supervision. Congress considered giving supervision of America's space program to the Department of Defense or—weirdly—to the Atomic Energy Commission. In the end, though, it decided to create a new agency on the bones of an already existing institution. The National Aeronautics and Space Administration (NASA, or N-A-S-A, as it was originally referred to) was established by the National Aeronautics and Space Act in July of 1958. Its official birthdate, 1 October, came just a few days short of the first anniversary of the Sputnik flight.

The new space agency inherited the research, employees, and budget of the National Advisory Committee for Aeronautics (NACA), a small but important government entity that did influential work in the areas of aeronautical design and testing. There was a sort of thematic resonance to this decision. The NACA was established during the First World War as an attempt to improve American aircraft production and performance. Despite the fact that Orville and Wilbur Wright invented the airplane, Europeans had in the years since leapfrogged American manufacturers, eventually creating powerful, highly maneuverable machines like the German Fokker D.VII and the British Sopwith Camel. Just so, and despite the fact that Robert Goddard invented the liquid-fueled rocket, NASA was now needed to the address the fact that the Soviets had somehow become world leaders in missile and space technology.

The NACA did valuable work, and it was partly responsible for the fact that American aircraft were equal or superior to enemy planes during the Second World War. Afterward, the agency oversaw development of both the x-1 aircraft, which was used by Chuck Yeager to break the sound barrier, and the x-15, which military and later NASA pilots flew to the lower limits of space. It was run by Hugh Dryden, a mathematics prodigy who graduated from Princeton University at the age of seventeen and helped create guided missile technology during the Second World War. While it employed a host of gifted theorists and technicians, perhaps NACA's most famous employees were Katherine Johnson, Mary W. Jackson, and Dorothy Vaughn, African American women who worked as "computers" and were commemorated years later in the book (and subsequent movie) *Hidden Figures.* Computers, as the name implies, were human number crunchers, capable of performing advanced calculations quickly and accurately. Jackson, it should be noted, eventually became NASA's first female engineer. Her name now adorns NASA's headquarters building in Washington DC.

Dryden became deputy director of NASA under the agency's new chief, Keith Glennan, formerly president of Case Institute of Technology (now Case Western Reserve University). It was a heady, hectic time. Glennan later wrote that "as I look back over my appointment schedules for those days, I wonder how

I kept anything straight." NASA was granted four existing research facilities (now known as the Langley, Ames, Glenn, and Armstrong Research Centers). It eventually assumed control from the army of both the Redstone Arsenal facility, in Huntsville, Alabama, which is now known as the Marshall Space Center, and the Jet Propulsion Laboratory in Pasadena, California, which is managed by Caltech University under contract with NASA. Perhaps the agency's most celebrated task at its inception was finding the human beings who would soon be asked, in the kindest way possible, to leave the planet. "What we're looking for here are just a few ordinary supermen," said a member of the selection committee. NASA reviewed the records and applications of the nation's best and brightest military pilots, subjected them to highly intimate and intrusive physical and psychological testing, and on 9 April 1959 introduced America's first seven ordinary super . . . er, *astronauts* to the world. More—much more—on these folks in a minute.

NASA also commissioned and began to build new facilities: the Manned Spacecraft Center (later Johnson Space Center) in Houston; Goddard Spaceflight Center in Greenbelt, Maryland; Stennis Space Center in Hancock County, Mississippi; and, perhaps most familiar of all, Kennedy Space Center on Merritt Island, Florida, site of the agency's crewed rocket launches. Long story short: NASA isn't a single facility or complex but rather a far-flung network of physical plants with different but sometimes overlapping functions. The various centers are of course meant to be complementary. However, there have occasionally been intra-agency rivalries, tensions, and turf battles. The straightlaced institutional types at headquarters, for example, haven't always vibed with the free-spirited robot builders at JPL; "spoiled brats," Keith Glennan once called them. The slow and steady rocket engineers at Marshall often rolled their eyes at the gung-ho astronauts and launch planners in Houston— who responded with exasperation at what they saw as excessive caution from the folks in Alabama.

There were good reasons for settling on the Sunshine State as NASA's number one launch site. First, the climate is famously mild. Second, Florida is relatively close to the equator, near where the earth is spinning fastest as it rotates. Because the earth rotates to the east, launching eastward from the Atlantic coast gives an ascending rocket a significant boost—as if an athlete were starting a fifty-yard dash by leaping off a carousel in the direction of the finish line. Furthermore, launching over the Atlantic provides a large unoccupied area where debris and even errant spacecraft can fall without hurting anyone below. And finally, Cape Canaveral had been used for years previously by the army and air force for rocketry exercises, so there was existing infrastructure—but little else—in the area.

As ambitious as the physical facilities and planning were, though, the new agency lacked a long-term goal. This was remedied by President John F. Kennedy. Kennedy visited Houston in September of 1962, shortly after the city was chosen as the site for the Manned Spacecraft Center. At Rice University, on an afternoon so hot and humid that even the birds seemed to sweat, the movie-star president repeated the remarkable challenge he'd issued to Congress a year and a half before. The United States, he said, would land a man on the moon and bring him safely back home by the end of the decade.

Kennedy was an unlikely space prophet. He had little interest in rocketry, which tended to be conducted by poorly dressed people in unglamorous places. Indeed, it was Kennedy's vice president, the beagle-slinging, elephant-eared Texan Lyndon B. Johnson, who rode herd on NASA from its inception through its victories in the midsixties. But the politics of the era left JFK no choice but to pay attention to the rapidly accelerating space race, and paying attention meant absorbing a rapid series of Soviet propaganda wins. He knew he had to do *something*, and the lunar landing had pop. It sizzled. So he seized it. And this afternoon in Houston, in one of the great American speeches of a war-torn century, Kennedy raised the challenge up over his head and brought it down hard. *To the moon*, he said! *By the end of the decade!* Using materials, the president casually added, that hadn't been INVENTED yet! There were smiles and cheers all around—even, it's rumored, a random yeehaw or two. The smiles on the faces of NASA administrators, however, were notably short-lived. Landing a man on the moon was the stuff of Saturday morning serials and flyboys in the Sunday papers. It was a daunting task. It was an outlandish ambition. In fact, some said it was impossible.

And suddenly the clock was ticking.

6

Shadows in the Sky

Created as a civilian agency, NASA inherited gifted people,
amazing machines, and extravagant plans from America's armed services.
But they didn't come without a fight.

The American space program began with the tinkering of a single civilian, Robert Goddard, who had no particular martial inclinations. By the late 1940s, though, in large part because of German advances during the Second World War, rocketry was a military project. Indeed, it was several military projects. The U.S. Army had Wernher von Braun, his team, and his scrapyard v-2s. These retooled "vengeance weapons" served as the basis for a number of America's early space vehicles, including the Redstone and Jupiter boosters. A Jupiter-C rocket sent America's first satellite, Explorer 1, into orbit on 31 January 1958.

Meanwhile, the navy worked with the Glenn L. Martin Company in the fifties to develop a v-2-based liquid-fueled rocket called Viking. Viking stood slightly taller than the v-2. Its shell was made of aluminum rather than steel, and its rocket motors could swivel, or "gimbal," which helped with control and maneuverability of the machine. The Viking rocket eventually became the basis for the Vanguard vehicle that launched America's *second* satellite into orbit.

The air force only became an independent branch of the armed services in 1947, when it was separated from the army. Nevertheless, the new service quickly initiated a number of missile and experimental aircraft programs, and by the midfifties, it had begun development of Thor and Jupiter medium-range ballistic missiles and their longer-range cousins, the Atlas and Titan weapons. Thus, each of the armed services had discrete and, in some respects, competing rocketry programs and increasingly ambitious plans for how their hardware could lead—or carry!—the nation into space. It wasn't just about the nozzles, propellant mixes, and thrust numbers. There were human lives, and human dreams, powering these visions. All through the early years of the Cold War, military personnel paved the way for the nation's eventual achievements in space, going higher, farther, and faster in an attempt to find the outer limits of both human physiology and high-tech machinery.

A few examples: Army Air Corps pilot Homer Boushey flew the first American rocket-powered aircraft in August of 1941. The craft was a small civilian plane equipped with Frank Malina's solid-fuel JATOs—essentially, giant roman candles that gave the plane a propulsive boost for rapid take-off. It was an Army Air Force pilot, Chuck Yeager, who broke the sound barrier in the rocket-powered x-1 (with "X" denoting experimental) aircraft in 1947—an important advance, technically and psychologically, since many people thought it couldn't be done.

On the ground, air force doctor and part-time maniac John P. Stapp strapped himself into a series of progressively more powerful solid-fuel rocket sleds to see how much stress from acceleration and deceleration he could stand. In his final test, in 1954, he survived 46 "g's," or forty-six times the force of "normal" gravity, which was two and a half times more force than the air force thought the human body could tolerate without undergoing a rapid unscheduled disassembly. During the test, he traveled briefly at a speed of 642 miles per hour before being jerked to a violent halt. He broke both his wrists, cracked some ribs, and struggled to breathe. Temporarily blinded by the effect of the stress pressing his eyeballs back into his brain and the resultant hemorrhaging, Stapp drew the line there—or possibly *there*. (It was hard to tell, because he couldn't see it.) When he later indicated he might be interested in trying to survive a sudden deceleration from a speed of *one thousand* miles per hour, his air force superiors pulled the plug on the project, unwilling to watch Stapp turn himself into Wile E. Coyote in pursuit of some chimerical psychic roadrunner.

In 1957 air force physician David Simons sealed himself in an aluminum cabinet and ascended via balloon to a height of 105,000 feet, far higher than any human being had gone before. He was participating in another of Stapp's programs, Project Man High, which was meant to determine whether human beings could function at high altitudes. It turned out that Simons could, at any rate, and the good doctor graced the 2 September 1957 cover of *Life* in what is thought to be the first selfie taken in the stratosphere.

Three years later, under the aegis of Project Excelsior, Captain Joe Kittinger rose skyward on a balloon-borne gondola launched from the New Mexico desert. He lost pressurization in his right glove as he ascended and had to watch as his unprotected hand swelled to twice its size, like the claw of a giant fiddler crab. At twenty miles up, and in excruciating pain, Kittinger tossed himself like a gum wrapper out of the gondola and fell eighty-four thousand feet toward a New Mexico desert. The air force pilot hit a top speed of 614 miles per hour—a record for unassisted human velocity for fifty-two years—on his way down before deploying his parachute for the final three miles. Unimpressed by this extended dance with mortality, Kittinger went on to fly almost five hundred combat missions in Vietnam before he was shot down by a North Vietnamese MiG-21 and imprisoned for eleven months in the facility known as the Hanoi Hilton. Kittinger was generous as well as valiant. Years after his release from prison and return to the United States, he helped Austrian daredevil Felix Baumgartner break the skydiving records he himself had set during his experimental air force work five decades earlier.

Naturally, NASA's initial space launches used military rockets and were sent up from military facilities. The first astronauts were military test pilots— not only in the familiar NASA Mercury capsules but also in sleek rocket-powered airplanes. The standout of these airplanes was the North American X-15, operated by the air force and NASA, which set speed and altitude records during the 1960s and reached the ragged edge of space. During the X-15 program, twelve pilots, including future NASA astronauts Neil Armstrong and Joe Engle, flew a combined 199 flights. Of these, eight pilots flew a combined thirteen flights that exceeded an altitude of fifty miles (80 km), thus qualifying as "astronauts" under U.S. Air Force criteria. Indeed, two flights by pilot Joe Walker exceeded 62.1 miles, the Kármán line, and thus qualified as spaceflights under international guidelines. The military pilots qualified for the Pentagon's version of astronaut wings immediately. The civilian pilots (including Armstrong) who accomplished the same feats were awarded NASA's astronaut wings in 2005, long after their qualifying spaceflights.

Epithets and Epaulets

Despite the considerable rocketry expertise and experience amassed by the armed forces, when Congress created NASA in 1958, it specified that the agency was to be controlled by civilians—a model strongly endorsed by both President Dwight D. Eisenhower and his vice president, Richard M. Nixon. This civilian branding of the agency was meant to reinforce the notion that the American space program was to be operated for peaceful purposes. Eisenhower also hoped that civilian control would mitigate inter-service rivalries over which bunch of uniforms would get to do what tasks in Earth orbit. This was sound thinking. Nevertheless, and despite congressional mandate, each of the armed services fought to create a role for itself in space operations, both through and apart from the new agency. One of the most urgent reasons for matching Soviet efforts, after all, was to ensure that the Russians didn't capture the strategic "high ground" of space. Air force brigadier general Homer Boushey—the same man who flew the first American rocket-powered airplane—gave a speech in 1958 in which he warned of the ability of a foreign power to launch nuclear weapons on the United States from the heavens. "He who controls the Moon," he advised, "controls the Earth." Von Braun had made a similar point in 1952, remarking that an orbital space station would be a "terribly effective atomic bomb carrier." If the American space program aimed to visit the moon, it was naturally seen by some in the Pentagon as an opportunity for occupation of that ghostly planetoid for military purposes.

In the wake of Sputnik, each of the armed services came up with its own plan for putting human beings into space. The army had Project Adam, which called for an astronaut to be sealed in a modified balloon gondola bereft of controls and launched on a ballistic trajectory through space, with an oceanic splashdown. The air force developed Project Man in Space Soonest (the unfortunately acronymed MISS) and even named Neil Armstrong to be one of the project's astronauts. The navy, meanwhile, cobbled together something it called Project MER-I, which referred to a "Manned Earth Reconnaissance" vehicle. While NASA eventually won control of crewed spaceflight operations, the overlapping goals and ambitions of the fledgling agency on the one hand and the objectives of the institutionally entrenched armed forces on the other led to some fierce competitions and awkward compromises along the way. One such conflict occurred early on, when NASA's congressionally directed acquisition of the army's Redstone Arsenal–based Ballistic Missile Agency (and its star engineer, von Braun) met with stiff but ultimately futile resistance from army brass. Around the same time, the service formulated what it called Project Horizon, a plan to build subterranean missile bases on the moon. Hopelessly beyond the nation's technological capabilities in that era, the blueprint for would-be moon warriors never attracted serious support in Congress. The air force, meanwhile, successfully campaigned to create and maintain a network of space-based spy satellites, working (and occasionally feuding) with the Central Intelligence Agency and later the National Reconnaissance Office.

The air force has always been the most strident and successful of the armed forces to lobby for a place in space. The army fights on land. The navy fights on the sea—and under it. For the nation's flyboys, it only made sense that the air force, which fights in the skies, should also develop the technologies to wage war *above* the skies. Thus, in 1957, the service developed a detailed and ambitious blueprint for its own crewed space program. Over the years it crafted plans for the futuristic Dyna-Soar orbital bomber independent of NASA, operated the highly successful x-15 rocket plane *with* NASA, and spent millions of dollars pushing the Manned Orbiting Laboratory Project (MOL), a secret space station concept studied seriously in the midsixties, in de facto competition with the civilian agency. The plan was to mate a modified Gemini capsule to a small cylindrical work station, launch this combined vehicle on a Titan III rocket, and then have its occupants return to Earth in the capsule when the mission was over. Like the Soviet Almaz space stations of the seventies, one of which was armed with a powerful cannon, the MOL was intended to be a military platform, used for reconnaissance (i.e., spying) on Soviet activities both on Earth and in low-Earth orbit. It might also have been used for defensive military and espionage operations and would, if produced, have ended up operating at the same time as NASA's *Skylab* missions. The air force spent considerable sums of money on the laboratory and even launched a test flight in 1966. Funding for the project was canceled by the Nixon administration in 1969, and seven of the air force's MOL astronauts were assigned to NASA. One, Dick Truly, eventually became NASA's chief administrator. Another, Bob Crippen, served as pilot on the first space shuttle mission.

Despite the setbacks represented by cancellation of projects like DynaSoar and the MOL, the air force never ceased its space-related activities or planning. As Chuck Yeager put it, "No blue suiter wanted to surrender space to NASA." Some of these plans met with failure. Project ABLE 1, sometimes referred to as "Explorer 0," was the air force's August 1958 attempt to launch a probe to orbit and photograph the moon. The mission ended only a minute after it started, when the probe's Thor booster rocket exploded in flight. Nevertheless, the service developed more reliable rockets and created its own remarkable history, lore, and set of firsts. In the late fifties, for example, the air force and the CIA collaborated to deploy the first of the Corona program's reconnaissance satellites. Here's where the "freedom of space" precedent set by the Soviets with Sputnik came back to haunt the USSR. Launched from California's Vandenberg Air Force Base on USAF Thor missiles, the Corona satellites orbited Earth from pole to pole at an altitude of between seventy-five and one hundred miles, photographing airfields, missile bases, and other areas of interest in the Soviet Union and Eastern Europe. The exposed film was robotically packaged and jettisoned from the satellite in what was called a "bucket." A small rocket slowed the bucket's momentum, which caused it to fall out of orbit. Once it hit Earth's atmosphere, the bucket deployed a large parachute. Then, if all went according to plan, an air force airplane armed with a giant hook scooped the payload out of the sky as it drifted downward and afterward ferried the film back to terra firma for processing and analysis. This midair catch was an intricate and awesome feat. The first one, executed in 1960, seemed miraculous. But the parachute scoops soon became routine. Though few Americans were aware of the fact, dozens of Corona satellites were sent up in the sixties, and the machines and their cameras became progressively more sophisticated. Altogether, the satellites took hundreds of thousands of surveillance photographs on some 2.1 million feet of film.

Despite the longevity and productivity of the program, Corona's work—indeed, its very existence—was only declassified in 1995. So a word of clarification. Though this narrative is a guide to the "American space program" writ large, many U.S. operations in Earth orbit have been and continue to be carried out in the name of national defense, and tenaciously hidden from public scrutiny. They are outside the scope of this history. Nevertheless, it's important to note that beside or behind NASA, the public face of American space exploration, a large, powerful, and mostly hidden realm of military space operations has evolved and is in fact more robustly funded and aggressively growing than the civilian agency.

Currently, the Pentagon's space program is dominated by the air force; its suddenly grown-up offspring, the U.S. Space Force; and the shadowy National Reconnaissance Office. Congress carved the space force out of the air force in 2019 and established it as a separate service, though it is still under the command of the secretary of the air force. The space force launches, maintains, and defends the nation's military and reconnaissance satellites and space-based information networks, including the Global Positioning System, or GPS. The fourteen thousand space force service people and civilian employees are known as Guardians. The force has its own uniforms, its own motto (*Semper Supra*, or "Always Above"), its own anthem, and—as frequently observed—a logo that looks a lot like the one used by James T. Kirk and his friends on *Star Trek*. Despite the Hollywood touches, the service is here to stay. Indeed, at $29 billion for fiscal year 2024, the space force has a budget significantly larger than NASA's, a reflection of the fact that space-based intelligence is now a vital component of the nation's warfighting abilities.

Perhaps no agency in government operates with as much secrecy as the National Reconnaissance Office, or NRO. Its existence can be traced to America's desperate attempts to understand the extent and potency of the Soviet Union's armed forces during the height of the Cold War. Such matters were closely guarded secrets, of course, and difficult to extract from a closed society in which both citizens and foreigners were closely watched for evidence of spying. An early attempt to probe Soviet nuclear capabilities involved the American release in 1956 of over five hundred surveillance balloons in Western Europe and Turkey. The balloons were meant to take advantage of prevailing wind currents to fly over Russia and other communist bloc nations and snap photos as they went, with recovery of the film to follow once the balloons had exited Soviet air space in the east. Project Genetrix didn't work out very well. Many of the balloons were lost or shot down by Russian fighter jets. Others failed to take photographs of any value. To make matters worse, the Soviets recovered some of the balloons and exhibited them to the public as evidence of Western perfidy.

Aircraft came next. The air force's famed U-2 spy plane performed good service over Russia until the Russians figured out how to shoot one down and capture its pilot, Francis Gary Powers, in 1960. Thereafter, the SR-71 Blackbird, an epically high-tech jet aircraft capable of flying at over three times the speed of sound at altitudes of some eighty-five thousand feet, took over. Such aircraft reconnaissance was helpful but limited in scope—and there was always a risk of confrontation and embarrassment in the court of world opinion. The first really revolutionary set of overhead images came from the Corona satellites, which delivered a treasure trove of photographs of Soviet airfields, missile installations, and other military bases. President Eisenhower was astonished by the yield. Not only did he want more satellite imagery—everyone else in Washington did as well.

Formed in 1961 to coordinate the satellite intelligence-gathering and dissemination efforts of the air force and the CIA, and formally controlled by the Department of Defense, the NRO was until 1992 so secret that the government refused to admit that it even *existed*. Employees refrained from saying its name, as if it were Voldemort, dark lord of evil forces. In the years after crewed space operations were taken away from the Pentagon, NRO became the yang to NASA's yin, a sort of reverse image of the familiar forms of the civilian space world. NRO watches the world but prefers to remain unseen while doing so. It spends massive amounts of money with little accountability. Indeed, at one point in the nineties, even after the agency had come in from the cold, Congress discovered that the NRO had managed to squirrel away some $2 billion in "rainy day" funds that no one had thought to account for.

The NRO commissions and orchestrates the operations of the nation's spy satellites, which circle the earth at various orbital inclinations and monitor and map every square foot of Russia, China, North Korea, and other potentially belligerent nations. While NASA looks up, NRO looks down. NASA publicizes. NRO keeps secrets. NASA has heroes: astronauts, engineers, and administrators, men and women whose names adorn buildings, book covers, and university brochures. NRO by contrast has villains: spies like Brian Regan, Christopher (the "Falcon") Boyce, and William Kampiles, whose names are seldom spoken. NASA has spaceships named after gods and explorers: Gemini, Magellan, Apollo, and Hubble. NRO has spacecraft named for random objects of no particular significance, as if our minds might just let these unremarkable monikers slip from memory: Hexagon, Onyx, Vortex, and Chalet. Some NRO programs are eventually declassified and thus made available for public scrutiny. Others remain hidden. The Corona program was made public in 1995, twenty-plus years after the program ended. A mysterious satellite known to amateur satellite trackers as Prowler, a possible surveillance platform that maneuvers close to and studies other satellites, was reportedly launched in 2010 but remains unacknowledged by NRO to this day.

And just as the "daytime" space program has books and movies reflecting its exploits, so too does this shadow empire—though admittedly not as many. One of the most noteworthy is *Ice Station Zebra*, the 1968 wannabe blockbuster that MGM shoved into theaters when it felt *2001: A Space Odyssey* had run its course. *Zebra* was nowhere near as good. And yet it has its charms. Inspired by actual events—namely, the disappearance of a Corona spy satellite's film "bucket" after it landed in Spitzbergen in 1959—the movie tells the story of a race between the Soviets and the West to recover film from a spy satellite that has crash-landed in the frozen wastes of the Arctic.

American spy satellites aren't alone in the heavens, of course. China, Russia, France, India, Israel, and an increasing number of other nations operate surveillance platforms in Earth orbit. Many of those platforms spend significant amounts of time studying the United States and American military assets. Under international law, and the precedent set by Sputnik, such flights are perfectly legitimate. The best thing that can be said for them is that they (a) decrease secrecy and the odds of a successful sneak attack by one nation against another and (b) increase the cost in people and machinery for a nation that initiates hostilities against a country with significant orbital surveillance and communication capabilities. They obviously haven't eliminated warfare, as any resident of Kiev will tell you. But they are not without their benefits either. Detection and deterrence have helped to keep the world relatively peaceful for the last fifty years.

Many Defense Department space operations have been kept confidential even when these activities were conducted with NASA personnel and/or assets. For example, ten space shuttle missions had military objectives, most notably the deployment of advanced reconnaissance satellites, and to this day remain classified—which means, basically, "not talked about." *At all*. At the conclusion of one such shuttle flight, STS-33, a DOD operative entered the orbiter and searched the astronauts' pockets to make sure no secret materials were going to leave the spacecraft.

As the United States faces challenges for control of Earth orbit from both China and Russia, the Pentagon's preference for secrecy will no doubt continue. The space force, for example, currently has a robotic "space plane," the X-37B, capable of flying in low-Earth orbit for years at a time. It's a spectacular accomplishment. While reconnaissance is its likely objective, the Pentagon refuses to say exactly what the vehicle does while aloft.

During the Cold War years, Soviet analysts often made the claim that NASA was a wolf in sheep's clothing—a military operation thinly disguised as a civilian project. This was never entirely true, but suspicions in this regard were understandable, especially given that the distinctions between the Soviet Union's own military and peacetime space operations were faint to nonexistent. And indeed there was substantial overlap in America's civilian and military space programs. As we've seen, NASA's early efforts used military personnel and hardware to accomplish ostensibly peaceful objectives. The companies that built the nation's lunar modules and space probes also produced our fighter jets and nuclear missiles. When NASA wanted to coordinate the vast and far-reaching efforts of its personnel and contractors, it "borrowed" General Sam Phillips of the air force to run the Apollo project. Gung ho air force lieutenant general James Abrahamson rode herd on the space shuttle program for a time in the mideighties, before being appointed by President Ronald Reagan to head the president's "Star Wars" laser-based defense system.

America's astronauts in the sixties and seventies were almost all military pilots, with ingrained service branch loyalties and biases. Even though they rarely wore their uniforms while working for NASA, most of the service-trained astronauts remained on active military duty. Naval officers serving in the astronaut corps were and still are required to undergo periodic sessions of "re-bluing," a sort of reorientation process in which they visit navy headquarters and attend presentations related to naval operations. While this helps navy astronauts keep abreast of developments at sea, it also helps the navy keep track of what's happening in space.

It wasn't until Harrison "Jack" Schmitt flew to the moon in 1972 on *Apollo 17* that a civilian who had never served in the nation's armed forces made it into space. These days there are lots of civilians in the program. Nevertheless, in NASA's most recent astronaut class, The Flies, in December of 2021, seven of the ten individuals selected were either active-duty or retired military officers, with another having served in the U.S. Coast Guard. So the military's influence on NASA continues. Its interest in the space program definitely continues.

But as for *control* of the program? Not so much.

7
Project Mercury

America's first crewed space program launched six astronauts on solo spaceflights during the two years from May 1961 to May 1963. The missions lasted longer each time, with the latter four accomplishing a total of thirty-four Earth orbits.

Sputnik was the best thing that ever happened to the American space program. It was a goad. A gadfly. It sparked a competition that led to amazing accomplishments by the engineers, designers, and astronauts of both the United States and the Soviet Union—as well as the consumption by the governments that employed them of huge amounts of manpower, money, and resources. The flight of the little satellite was just the first blow. The Soviets followed up quickly on their success by launching Sputnik 2. The USSR then became the first nation to reach the moon when it landed its *Luna 2* probe on the lunar surface in 1959. *Landed* is one way to put it, anyway. The probe hit the lunar surface at several thousand miles per hour, destroying itself and scattering a collection of titanium-alloy trinkets stamped with Soviet emblems. In October of 1959 *Luna 3* took photographs of the far side of the moon. The moon is tidally locked with Earth, which means that we always see one side of it. The Soviet images revealed that the sphere's far side (sometimes called its "dark side," though it does in fact receive sunlight) is far more pockmarked and cratered than the near.

These socialist exploits stung NASA, which was quite literally having trouble getting the U.S. space effort off the ground. After the disastrous failure of the Vanguard TV-3 mission in December of 1957, Wernher von Braun's army team launched America's first satellite, Explorer 1, the following month. The navy's follow-up effort, Vanguard 1, made it into orbit in March of 1958, becoming the world's first *solar-powered* satellite. So America could launch satellites too—on a good day, at least. There were numerous other failures along the way. But everyone knew satellites were just a prelude to the main event: putting a man in space. And creating a spaceship fit for human occupancy was a daunting task, with a long list of potential hazards. The new vehicle would need to protect its astronaut from the airless vacuum, extreme temperature fluctuations, harmful radiation, and jagged micrometeoroids of space. It had to shield him from the intense heat and pressure of reentry into Earth's atmosphere and cushion his impact with either ground or water at the conclusion of the flight. And this was just the start. This was the minimum. The new spacecraft required controls and monitors, radio equipment, a launchpad escape system, parachutes, and an inflatable cushion to float on in the choppy waters of the Atlantic after splashdown.

Even the *shape* had to be invented. The notion of a teardrop-shaped capsule—a "truncated cone," as one source calls it, "shaped roughly like a television picture tube," or "cone-shaped, with a neck at the narrow end," as someone else saw it—narrow at the top and broad at the bottom, seems commonsensical to us now. But the configuration wasn't inevitable. One astronaut commented that the capsule as originally envisioned—before it became a truncated cone, that is—would have looked like an upside-down coffee cup perched on top of a rocket. The Soviets by contrast encased their cosmonauts in spherical reentry vessels for many years, with the design changing eventually into something that more closely resembles a champagne cork.

Design honors for America's first spaceships go to Max Faget, NASA's no. 1 toymaker to the king. Faget was the son and grandson of pioneering physicians—his father, a tropical disease specialist, developed the first drug treatment for leprosy. Faget studied engineering at Louisiana State University, served as a submarine officer during the Second World War, and joined NACA as an engineer in 1948. He was a slight man who looked a bit like the actor Ray Walston, with wide-set eyes, prominent ears, and a hairline that began an orderly tactical retreat from his forehead early on in his life. Elfin and energetic, he was known to do handstands when faced with a difficult problem in order to stimulate his thoughts. With NACA, he had a hand in the creation of the x-15 space plane and a variety of intercontinental ballistic missiles. As a NASA employee, he invented form-fitting fiberglass acceleration couches for the astronauts, molded to the contours of their bodies to lessen the strain of high g forces. He sketched out the emergency ejection tower for early launch systems and is generally credited as the lead designer of the Mercury, Gemini, and Apollo crew capsules. These capsules are narrow at the top to fit the streamlined profile of a rocket designed to punch a hole in the atmosphere on the way into space. They are broad and rounded on the bottom for a number of reasons. First, the capsule is aerodynamically stable when falling blunt-end first. Furthermore, the larger blunt end provides for increased drag to slow the space vehicle and for maximum deflection of heat as the capsule reenters Earth's atmosphere. Faget deserves more recognition. Publicity shy and elusive, he is about as obscure as anyone called the Father of the American Space Program could be.

The Mercury capsule was constructed of a nickel alloy called René 41, with an outer shell of titanium plates. Its blunt bottom was buttressed by an "ablative" heat shield, a layer of materials specifically designed to absorb heat and gradually vaporize, taking the heat along with them. The Mercury capsule's rounded bottom also provided a modest amount of lift in the proper attitude, giving the astronauts a degree of maneuverability during reentry.

That the astronauts could maneuver at all was another detail that had to be worked out before Mercury's first launch. NASA originally wanted the capsule's flight to be fully automated, with the astronaut himself just an observer: "Spam in a can," as wags of the day put it. This plan met with staunch resistance from the astronauts themselves, who fought for and won the ability to actually "fly" the spacecraft—to the extent the capsule could fly at all. Once in orbit, the astronauts were able to maneuver their vehicle along its longitudinal axis (roll), left to right from the astronaut's point of view (yaw), and up or down (pitch), with movement created by hydrogen peroxide–fueled thrusters. They won other concessions as well: a window, for one, and a hatch, or door, they could open themselves after splashdown. They also lobbied to have the term for the vessel they'd occupy in orbit changed from *capsule* to *spacecraft*, to reflect that the astronauts weren't just occupants of the machine but pilots of it. This, too, was accomplished, at least in official circles. The public, meanwhile, still used the term *capsule*, as if the space vehicle were a medicine the nation was taking to restore its public pride.

But it wasn't just spacecraft that had to be imagined, invented, and, well, *named*. NASA had to establish radio communications stations around the world—on land in Australia and Mexico, Zanzibar and Bermuda but also at sea, on specially equipped vessels—in order to exchange information with its spaceships as they flew overhead. The agency needed protocols and checklists. It required spacesuits, helmets, gloves, and boots. It had to have some way for personnel on the ground to share information with each other as the situation in the sky evolved. As Chris Kraft, the dagger-sharp Virginian who invented what we now call "mission control," remembered those early days, "We didn't have any buildings, we didn't have any radar, we didn't have any telemetry, we didn't have any voice communications, and we ended up then saying, these are our requirements. We gotta tell 'em it's gotta have a computer. *What the hell is a computer?* It was about that much that we didn't know." NASA borrowed one important element of the launch process, the countdown to zero, from expressionist German cinema, as the reverse numbering was evidently first used in Fritz Lang's 1929 sci-fi flick *Woman in the Moon*—technical adviser, Hermann Oberth.

The infant agency craved numbers: orbital calculations, fuel burn projections, launch and reentry parameters. It gobbled up money, so Congress sent more. NASA's budget increased 700 percent in its first four years of existence and continued to climb steeply in the following four-year period, reaching a high of some 4.41 percent of the federal budget in 1966. (It's now less than half of 1 percent.) NASA went from eight thousand employees at its creation in 1958 to thirty-six thousand in the same period of time.

Slowly, the American genius for logistics, so crucial to the country's military success in the Second World War, began to assert itself. Engineers, draftsmen, and data processors moved to new digs in Cape Canaveral, Huntsville, and Langley, Virginia. The agency acquired and installed walls full of IBM 7094 computers, eerie harbingers of a mechanized future with massive 150-kilobyte memories and twin spools of magnetic tape mounted like the eyes of observant owls. Whole new suburbs sprouted in the cow pastures southeast of Houston, where NASA's new Manned Spacecraft Center was under construction. Defense contractors—Lockheed and Grumman, Martin and North American, to name a few—retooled and ramped up. The future was frantically inventing itself. Only problem was, the future seemed to speak Russian.

On 12 April 1961 a diminutive Russian fighter pilot named Yuri Gagarin was bolted into a giant bowling ball called *Vostok* that was attached to an R-7 missile. The missile was aimed at the sky. This being the Soviet space program, a secret code that would enable Gagarin to take manual control of the secret spacecraft was hidden in a secret place in the cockpit. A communist functionary whispered the code in his ear—*secretly*, of course.

The engines rumbled. The ground shook.

"Here we go!" said Gagarin.

And off he went.

America's First Flight

In the early years, the Soviet space program had one major advantage over its American counterpart: No one could see it. Only a select few Soviet functionaries knew that the Russians were going to launch a rocket until it was already up. Thus, it always appeared that the Soviets knew exactly what they were doing after they did it—or, in some cases, lied about doing it. By contrast, NASA was plagued not only by mistakes and mishaps but also by the fact that its failures were broadcast on the nightly news, fodder for newspaper hacks and professional comedians. The Vanguard TV-3 explosion spawned a cascade of "kaputnik" jokes in 1957. The test flight of Mercury-Redstone 1 in November of 1960 made it less than a foot off the launchpad before the Redstone's engine shut off and the rocket settled back on the pad, fully pressurized but now unmoored, at risk of falling over, and—like Vanguard—erupting in flame. Mission director Kraft thought about sending someone out with a gun to shoot it. Not as an expression of either mercy or madness, but rather to puncture the Redstone's pressurized propellant tanks so the whole thing wouldn't go off like a giant cherry bomb. Calmer heads prevailed, though, and the rocket's tanks eventually bled off enough fuel to depressurize themselves. More media hilarity ensued. Nevertheless, the affair caused anxiety in the astronauts and heartburn in the heartland and earned Mercury-Redstone 1 the epithet "The Four-Inch Flight."

But the agency learned. Despite some rough initial outings, things started to click for NASA with the Mercury program. Named for the wing-footed Roman god, nimble intermediary between the whispering realms of the dead—that is, the past—and the clamorous world of the living, Mercury was a series of increasingly ambitious solo spaceflights featuring six of the fabled Original Seven astronauts.

The boyish but cold-eyed Alan Shepard gave the project its organizing credo as he sat on the launchpad, waiting to become the first American in space. Shepard was capable of camaraderie one minute and contempt the next. His broad grin exposed a pair of extended canines that looked a little like fangs, as if he'd been both a lieutenant and a lycanthrope. It was 5 May 1961. After four hours of immobility in a stiflingly hot metallic spacesuit with twenty-seven zippers, the former naval aviator was venturing out beyond the planetary limits of courtesy. NASA's slender white spaceship, capped in black, stood in the sunshine like the turret of a Bavarian castle. The world was watching, but delay followed delay. It was embarrassing. The Soviets could do it. Why couldn't we? NASA's engineer army, that bespectacled, white-shirted corps of number crunchers and slide-rule jockeys, fretted over every detail. Shepard grew exasperated. "Why don't you fix your little problem," he radioed to mission control, "and light this candle?"

Thirty-six thousand pounds of liquid explosive duly lit, Shepard rode a Redstone rocket into space and quickly back down again, smacking into the Atlantic Ocean not far from the Bahamas. He experienced weightlessness shortly after the Redstone's motors shut off, around two and half minutes after launch, but was able to take manual control of the capsule and found that changing the spacecraft's orientation was easily manageable. For purposes of international bragging rights, it's quite correct to call Shepard's accomplishment the first *piloted* spaceflight, as Gagarin's sojourn was controlled from the ground. Mostly controlled, anyway. Because the Vostok capsule had no "braking" apparatus that would have allowed for a survivable landing, Gagarin was ejected from his spacecraft at an altitude of twenty-three thousand feet and parachuted to Earth, where he landed near the western border of Kazakhstan.

John Glenn Orbits Earth

Yuri Gagarin's hop into the heavens handed the Soviet space program a propaganda coup almost as big as the launch of Sputnik. "MAN ENTERS SPACE" headlined the *Huntsville Times*, adding what scarcely needed to be said: "U.S. HAD HOPED FOR OWN LAUNCH." Shepard's mission a month later seemed underwhelming by comparison. Gagarin, after all, had orbited the earth; Shepard simply flew into space and fell back down again. Shepard won some style points by making a more or less controlled landing in the ocean, whereas Gagarin was unceremoniously ejected from his Vostok craft at four miles up, but the Soviets kept that part of the mission under wraps. Nevertheless, the first American manned mission was a hit. When Shepard's Mercury capsule went on public display in France, some six hundred thousand Parisians lined up to view and, in many cases, stroke the charred spaceship.

Energized by its early lead in the court of global opinion, Moscow hurried to put up additional space spectaculars in the early sixties. Cosmonaut Gherman Titov became the first man to make multiple orbits around Earth (seventeen, in fact) when he flew on *Vostok 2* in August of 1961. He also took the first moving pictures from space and became the first person to sleep while

PROJECT MERCURY | 71

in orbit. Not all of these successes actually advanced the state of Soviet aerospace engineering. Sending two cosmonauts into space at the same time, as the Kremlin did in August of 1962 made sense; Russian moon landing plans, preliminary as they were, involved a two-person crew, and everyone agreed that it would be important for one spaceship to be able to rendezvous with another. But launching a Voskhod vehicle in October of 1964 with *three* men crammed inside was pointless, more a stunt than a step. (Space historian Colin Burgess notes that even some Soviet observers called the flight a "fiasco" and a "space circus.")

The United States would continue to lag behind the Soviets for years. But there were signs early on that the race wasn't over. On 20 February 1962 Marine test pilot and astronaut John Glenn, rock-ribbed and guileless as a Presbyterian hymn, climbed into his Mercury capsule on a Cape Canaveral launchpad to embark on the nation's first orbital spaceflight. The launch was a gut check for NASA, as orbital flight is considerably more difficult to pull off than a mere up-and-down ballistic trajectory like that of Al Shepard in the first Mercury mission or Gus Grissom in the second. To achieve orbit, it's not enough for a rocket to boost its payload—whether dangerous warhead or glamorous jarhead—above the Kármán line. Rather, the rocket has to put the payload into space at a sufficient velocity to keep it there. To do this, the rocket has to go fast. *Really* fast. At too low a velocity, John Glenn falls back to Earth. This is bad. On the other hand, at too high a velocity, *Friendship 7*, Glenn's capsule, soars out into space, never to return. This is worse. The aim is to put the capsule into the cosmos on the right flight path and velocity to keep the spacecraft falling around the planet at a more or less constant speed relative to Earth, held in its course by the pull of Earth's gravity on the one hand and its own inertia on the other. This velocity is commonly said to be 17,500 miles per hour. It's an approximation, as the velocity needed to maintain orbit varies depending on how far away from the planet the spacecraft is. But 17,500 miles per hour—slower than a lightning bolt, faster than a bullet—is the generally accepted shorthand for orbital velocity.

And everyone knew it was *doable*. Heck, the Russians had done it with Sputnik five years earlier, and then with Yuri Gagarin crammed inside a spherical ship back in April of '61. But now it was time for an American to try, and no one familiar with the sad panoply of early American space launches was feeling entirely optimistic. Between fits of obsessive plotting to overthrow Fidel

Castro, the Pentagon made plans to blame any mission failure on Cuba. Nevertheless, thousands of spectators gathered on Florida beaches to watch what a Tampa newspaper called—not entirely reassuringly—a "manshot." Their worries quickly dissipated. Glenn took off in a blaze of fire and smoke and circled the globe three times before guiding his little capsule back through the atmosphere to a landing in the Atlantic Ocean. His photogenic chin seemed to emerge from *Friendship 7* a full second before the rest of him did. "SPIRITS OF AMERICANS SOAR AS HIGH AS SPACEMAN GLENN" reported the *Vancouver (BC) Sun*.

The Mercury program's two rockets were modestly sized by modern standards. The Redstone, for example, stood just over 80 feet tall and weighed 66,000 pounds. By contrast, the Saturn V, which was eventually used to send American astronauts to the moon, stood 363 feet tall and weighed 6.2 *million* pounds. The Redstone delivered 78,000 pounds of thrust; Saturn V, some 7.5 million pounds. Nevertheless, the Redstone was powerful enough for the first two crewed Mercury flights, each of which was suborbital. It was replaced for the next four orbital missions by the Atlas. The Atlas rockets were, in fact, modified intercontinental ballistic missiles, with men instead of warheads bolted onto the top. The Mercury capsule was also quite small: ten feet high, not counting its escape tower, six feet in diameter on the blunt end, only two and a half feet at the top, with thirty-six cubic feet of living space inside. Outfitted in his pressure suit, noted *Life* magazine, "the astronaut fills the thing completely to its limit. There is room for nothing else." Cutaway illustrations of an astronaut nestled in his capsule seem to show a man driving a lightbulb, facing away from the filament. It was a tiny conveyance, porcupined inside with switches, levers, warning lights and circuit breakers. And wires: the little capsule reportedly held seven *miles* worth of electrical wiring. It had just about the same air volume, wrote historian William E. Burrows, as "a large adult casket."

Modest as they were in retrospect, the Mercury missions sufficed to create the B-roll of the American space program: silver-clad astronauts like modern-day knights in some peculiarly Yankee combination of plate armor and chewing gum foil, dramatic countdowns and fiery launches, distant splashdowns on gorgeous seas and the sight of grizzled spacefarers waving from the deck of a highway-colored aircraft carrier. On the sixth and final Mercury mission, in May 1963, Gordon ("Gordo") Cooper was launched on an Atlas rocket into low-Earth orbit and circled the planet twenty-two times over the course of thirty-four hours.

A legendarily laid-back and occasionally unpredictable pilot, the youngest of the Original Seven, the good-natured Oklahoman dozed off on the pad while awaiting launch. He slept again while in space. When electrical problems in the capsule required him to take manual control of the craft toward the end of the flight, Cooper brought it back to Earth using dead reckoning and good old cowboy stick-and-rudder skills. It was the perfect blend of technological accomplishment and aw-shucks American bravado, Audie Murphy and Annie Oakley, Lucky Lindy and Orville Wright. Suddenly NASA was on a roll. America could put human beings in space.

Now it was time to find out what they could do once they got there.

Eleven Boffo Space Books to Launch at Your Brain

1. Michael Collins, *Carrying the Fire*: Generally considered to be the best of the many astronaut biographies, Collins's memoir is a joy to read—intelligent, humane, and candid.

2. Andrew Chaikin, *A Man on the Moon*: Another gold standard. Andy Chaikin's big book is the go-to history of the Apollo program.

3. James R. Hansen, *First Man*: This exhaustively researched biography of Neil Armstrong will answer all your questions about America's secular saint except the most important one: *Why can't we all be him?*

4. William E. Burrows, *This New Ocean*: Burrows's history of space exploration came out in 1999, which means it's dated now. But for the story up to that point? Wow! The author writes knowingly (and cleverly) about just about everything, offering a sort of narrative encyclopedia of humanity's first climbs into the cosmos.

5. Andy Weir, *Project Hail Mary*: It's hard to know exactly what to make of this expansive and elaborate imagining of humanity's first contact with a rock-like alien species that desperately needs our help. There's occasionally a little too much detail in *Project Hail Mary*, but it's balanced out by Weir's second-gen humor (he's much funnier here than in *The Martian*) and a surprisingly touching story about two wildly different types of science nerd who end up forging a world-saving alliance. *Amaze.*

6. Al Worden and Francis French, *Falling to Earth*: In the midst of a sea of big and self-important bios, Worden's story is a quiet reminder of the humanity of NASA's spacefarers, written by an *Apollo 15* astronaut who lost his way and haltingly found a path back.

7. Mike Mullane, *Riding Rockets*: Shuttle astronaut Mullane's sense of humor isn't for everyone, but *Riding Rockets* is a valuable look at the "new NASA" ushered in by the selection of the 1978 astronaut group, The Thirty-Five New Guys, which (finally) included women and African Americans, Asian Americans, and Jewish Americans.

8. Norman Mailer, *Of a Fire on the Moon*: As the world watched, America's most assiduously narcissistic man of letters tried to write the definitive account of the "meaning" of the *Apollo 11* moon landing. While Mailer by his own admission never quite figures it out, watching him stalk the likes of Armstrong and von Braun in search of some gnomic truth beyond the scrubbed and odorless WASP culture of NASA is worth the price of admission.

9. Michael Neufeld, *Von Braun: Dreamer of Space, Engineer of War*: Neufeld travels far, far upriver to study NASA's heart of darkness, the brilliant, protean Prussian nobleman whose work shaped the German, American, and Soviet space programs. There's the glory, of course. And the horror.

10. Tom Wolfe, *The Right Stuff*: Despite all the hyperbole in the book and the cartoonish movie that was made of it, Wolfe's tome was a serious attempt to both humanize and lionize the Original Seven in the face of sometimes condescending popular criticism.

11. Stephen Walker, *Beyond*: We're still learning about the secretive Soviet space program, which for decades inspired fear, envy, and admiration in American public opinion and NASA's planning meetings. There's no better place to start your journey to Star City—Russia's equivalent of the Johnson Space Center—than with this brilliantly paced, fact-filled portrait of Yuri Gagarin, the first person to reach the cosmos.

8

Gemini's Forgotten Flights

In the Gemini program, NASA sent up ten crewed spaceflights between March 1965 and November 1966. Each flight carried two astronauts, who were assigned to complete progressively more complicated tasks meant to demonstrate the viability of a lunar landing mission. No one remembers anything about these events. No one, that is, except us.

You don't need money to become a space nerd. You don't have to sign up for classes or know secret code words or submit to a demeaning selection process by NASA's fabled "sorting helmet." But you do have to get your Gemini missions straight, because that's how we're able to identify each other at wedding receptions and Super Bowl parties. Only speak the names and mission numbers, friend, and join the cognoscenti. We'll be waiting on the patio. Bring the queso.

But first things first. After Mercury came Project Gemini, which—*insider tip*—astronauts of the era always called "Geminee," to rhyme with Walt Disney's animated cricket, Jiminy, as if the whole enterprise were a talking space bug. Each of the Gemini program's missions was crewed by two astronauts rather than one—hence the name, which refers to Castor and Pollux, the immortal twins of Greek mythology. Pollux was the son of Zeus (in the form of a swan, supposedly) and a beautiful princess named Leda, while Castor was the son of Leda and the Spartan king Tyndareus. Pollux loved his half-brother so much that he implored Zeus to grant him immortality, just as Pollux enjoyed. The god's solution was to transform the siblings into the two brightest of the stars that form the Gemini constellation. Castor and Pollux are considered to be patrons and protectors of travelers, especially sailors, to whom they appear as St. Elmo's fire—not the cringeworthy 1985 movie, mind you, but the eerie electrical phenomenon.

Project Gemini started with a test flight, *Gemini 1*, in April of 1964. The first *crewed* flight, *Gemini 3*, was launched on 23 March 1965, and the series of missions continued until the return to Earth of *Gemini 12* nineteen months later. It was an eventful year and a half. The flights were launched on Titan II rockets, which, like Mercury's Redstone and Atlas launch vehicles, were adapted from their original purpose of carrying explosive payloads. The new rockets were needed to boost a larger, heavier capsule. Technically known as the "reentry module," the Gemini capsule was a larger version of the Mercury vessel—though only 50 percent bigger inside, despite the addition of a second crewmember. However, the Gemini capsule was wedded to a truncated-cone-shaped "adapter" module that housed oxygen tanks, communications equipment, and the thrusters and propellant that allowed the crew to fly their spacecraft while in orbit. When considered as a whole—that is, reentry and adapter module together—the spacecraft was ten feet wide at the base and eighteen feet tall and weighed around 3,500 pounds, depending on the mission. Less momentous than the first Mercury flights but equally important, the Gemini flights focused on achieving some of the operational tasks, like rendezvous and docking with other spacecraft, that missions to the moon would have to accomplish.

The Gemini capsule also had two hatches, one above each astronaut, that could be manually opened and closed while in orbit. This improvement over the Mercury capsule allowed for the first American space walk. Opening a hatch resulted, of course, in a rapid and complete loss of pressure in the spaceship, as the pure oxygen in the cabin rushed out into space. We think of pressure like it's a bad thing. We're under pressure to perform. We spend the weekend depressurizing. Quit *pressuring* me, we say—I'm not going to buy any more cryptocurrency! But life without pressure is no life at all. Take space. One of the challenges associated with traveling through the cosmos is that there's no pressure in space. Any human being exposed to the vacuum of the void would feel excruciating pain as his or her blood began to bubble, the oxygen no longer imprisoned in the astronaut's viscous vital fluid but now sizzling and popping in every vein, as if one were being tased from the *inside*. (We don't have many first-hand reports regarding depressurization of a human subject, but one person who lived to tell the tale was Jim LeBlanc. In 1966 LeBlanc was assigned to test a NASA pressure suit in a vacuum chamber. The suit suddenly lost pressure. LeBlanc passed out shortly afterward, but he was pulled to safety by colleagues before he died. He remembered feeling, as writer Paul Parsons puts it, the "saliva boiling on his tongue.") The temperature up here might be 250 degrees or negative 250 degrees, depending on whether our astronaut is exposed to direct sunlight. A spacefarer stranded even five feet away from the shelter of a spacecraft would have no way to return to it without some external aid—a push, a shove, a lifeline or engine—as the heavens offer no purchase for walking, crawling, or swimming. And there would be no air to breathe, as there is no atmosphere in space. Hence the pressure suit, a sealed metal-and-fabric cocoon containing its own little biosphere, plus temperature controls, and—because of the pumped-in atmosphere—*pressure*. Flying is hard, certainly. But flying in space is harder. That's why astronauts wear armor when they step into that big black sea.

Without it, they'd be dead.

A Stroll in the Stars

The world's first spacewalker was Russian. Short and stocky, quick with a grin, Alexei Leonov was the most approachable of the Soviet-era cosmonauts. Leonov left the cramped confines of his Voskhod (sunrise) spacecraft in March of 1965 and floated beside it for twelve minutes, firmly moored to the Voskhod by an eighteen-foot metallic tether. This was the world's first in-space extravehicular activity—the first EVA, as NASA calls it, or "space walk," as the rest of us do. Leonov noted how quiet space was. He could hear his heart beating. And he marveled at his own tininess in the jaws of forever. "My feeling," he said, "was that I was a grain of sand."

The cosmonaut had trouble getting back into the Voskhod. Like a bag of potato chips brought from the lowlands to the mountains, his pressure suit puffed up so much in the vacuum of space that Leonov got stuck in the spaceship's inflatable airlock. The Russian had to depressurize—that is, let the air out of—his suit, reducing the oxygen in it to dangerously low levels before he could squeeze through the airlock. At this point, he said, sweat was "sloshing" around inside the suit with him. His heart was racing, and he was close to collapse. He might have died. It was that close. In announcing Leonov's successful space walk, however, the Soviets neglected to mention the difficulties involved. They played up Leonov's feat as simple and easy, a stroll in the cosmos for New Socialist Man.

The space walk was another propaganda win, one of many for the Soviets in those days, and NASA felt compelled to respond in kind. On 3 June 1965 American astronaut Ed White ventured outside his Gemini capsule with the means not only to float but to *fly*, gripping a gas-powered gizmo nowadays referred to as the "hand-held maneuvering unit," or HHMU. The little wand wasn't very helpful. It ran out of its compressed oxygen propellant at about minute three of White's walk, leaving him to flail around under his own power through pulling on his tether.

Despite his depleted HHMU, White enjoyed the experience of weightless movement so much that he had to be urged back inside the spacecraft. "I'm coming back in," he said, as he returned to the capsule, "and it's the saddest moment of my life." Mission commander Jim McDivitt's photographs of his crewmate floating a hundred miles above Earth, connected to the mothership by only a slender golden cable, quickly became emblematic of the American space program. The race for preeminence in the cosmos was partly about perception. While White wasn't the first spacewalker, the images of his feat were far more interesting than those taken of his Soviet counterpart, Leonov. The Gemini photos, vivid, crisp, and inherently dramatic, allowed the United States to pull even in the public relations race with Moscow. While the Russians might be first to arrive at some important destination, it was the Americans who were sending back postcards.

Why We Float

Weightlessness in space is often attributed to a lack of gravity. This is inaccurate. By definition, any object orbiting Earth is subject to the planet's pull. This is why it's in *orbit*, rather than traveling out toward distant stars. What Ed White felt as he circled the planet at 17,500 miles per hour wasn't the absence of gravity but rather the sensation of falling—falling endlessly but never descending because his velocity was so high that he continued to circle the planet, as his inertia and the bearlike embrace of the planet's gravity kept him at the same altitude as he traveled. He would *eventually* have slowed down due to the very small but still measurable drag created by Earth's atmosphere at that altitude, just as a space station eventually slows down if not boosted periodically by thrust from a rocket booster. But as long as White's velocity was maintained, he would have fallen forever—or for as long as the planet is around, which, for most purposes, is long enough.

The Missions

Gemini 4 was the first mission to be controlled by personnel in NASA's new mission control center in Houston rather than at Cape Canaveral. *Gemini 5* lasted just over a week and set a record (soon broken) for spaceflight duration—an important expansion of the envelope at a time when scientists weren't sure how the human body would function during extended periods of weightlessness. *Gemini 6* and *Gemini 7* rendezvoused ("met up") in orbit on 15 December 1965, 150 miles above Earth, the first vehicles to perform this important task. On *Gemini 8*, astronauts Neil Armstrong and Dave Scott rendezvoused and docked their capsule with a remote-controlled Agena spacecraft in one of a number of exercises designed to test procedures that would later be needed for a lunar mission. The docking went fine. Shortly afterward, however, a thruster stuck in the open position and pushed the capsule and its two occupants into a combined spin and tumble.

Thinking that the problem was caused by a thruster on the Agena, mission commander Armstrong disengaged the Gemini capsule from the drone. It didn't help. In fact, disengagement made the problem worse. There was an open thruster, but it was a *capsule* thruster, and the decreased mass of the capsule now that it was detached from the Agena allowed the tumble rate to increase. The Gemini's movement became so violent that it almost rendered the two astronauts unconscious. This would have led to their deaths. Disoriented and barely able to see the instrument panel, Armstrong managed to turn off the main thruster system and activate the spacecraft's reentry thrusters, which eventually brought the capsule back under control. In the program's first orbital mission abort, NASA ordered the astronauts to return to Earth, as was required by mission rules whenever the reentry thrusters were fired.

Armstrong's actions doubtless saved his own life and that of Dave Scott and helped establish Armstrong's reputation as a man whose blood turned to radiator coolant under pressure. "It was my lucky day to be flying with him," Scott later said. It was a bit of an understatement, all things considered.

Gemini 9 was star-crossed from the start. Its original crew, Elliot See and Charlie Bassett, died in a crash of their T-38 aircraft in St. Louis on 28 February 1966, so backups Tom Stafford and Gene Cernan took their places. Once in orbit, their attempt to dock with another Agena failed when the Agena's nose shroud failed to detach. After that, Gene Cernan attempted to test a hydrogen peroxide–powered jetpack called the astronaut maneuvering unit. The jetpack was originally an air force idea and another effort to liberate astronauts from their tethers. The test was a bust. In fact, Cernan so exhausted himself during his two-hour attempt to get to and put on the jetpack—the

"Space Walk from Hell," he called it—that the experiment was called off, and Cernan was directed to return to the capsule.

Gemini 11, crewed by Pete Conrad and Dick Gordon, set a record for Earth-orbit flight altitude at 853 miles and executed a series of important rendezvous and docking procedures with an Agena target vehicle, including docking with Agena on *Gemini 11*'s first orbit. Gordon's initial EVA on the mission had to be cut short because he, like Cernan, became exhausted by working outside the capsule, but he did manage to accomplish his chief objective, which was to attach a tether from the *Gemini 11* capsule to the Agena target for use in a series of studies of how the two spacecrafts' movements affected the other. Conrad managed to put the tethered vehicles in a controlled spin and thereby generated a small but measurable amount of artificial gravity as a result of centrifugal force. After four days in orbit, Conrad relinquished control of the spacecraft, and it reentered Earth orbit on auto-pilot, splashing down just six miles from its retrieval ship, the USS *Guam*.

The final Gemini flight, *Gemini 12*, took place in November 1966 and featured three EVAs by the hyper-focused rookie Buzz Aldrin, who took photographs, retrieved a micrometeoroid collection experiment attached to the exterior of the capsule, and generally demonstrated that astronauts could work outside a spacecraft in orbit, given the right hardware, a modest task list, and proper preparation. *Gemini 12* was commanded by the genial Jim Lovell, who would later fly on the *Apollo 8* and *Apollo 13* missions, both of which flights were miraculous, though for different reasons.

Aiming Higher

If one were asked to choose a year or two from its history as the best time in which to live in the United States, the midsixties would be a smart pick. Not for everyone, of course—and not in every regard. But the nation's majority population prospered during this era, which can be seen as the cresting wave of WASP culture. Thanks to NASA's series of ten two-man Gemini flights, the humiliations of the early space race were over. The Cold War seemed to have stabilized, with neither side holding much of an advantage over the other. African Americans were marching to protest segregation and race hatred in the South, and LBJ was steadily increasing the U.S. presence in Vietnam, but the country still thought of itself as prosperous and peaceful.

Americans were already the wealthiest people in the world by a sizeable margin, and our standard of living continued to climb. Cars got faster. Homes got bigger. Children were healthy, schools were good, and the world seemed like a manageable place. *Mary Poppins* and *My Fair Lady* were big winners at the 1965 Academy Awards; *The Sound of Music* triumphed the following year. Mouth-breathing fare like *Bonanza* and *Gomer Pyle, U.S.M.C.* flickered on our TV sets. The Beatles played Shea Stadium to the sounds of screams and shrieks, but it was the lighter-than-air British quintet Herman's Hermits who dominated the pop charts, spinning melodies like cotton candy and registering five of the Billboard Top 100 hits of 1965, including the nebbish anthem "Mrs. Brown, You've Got a Lovely Daughter." The Broadway production of *Man of La Mancha* introduced Americans to a song called "The Impossible Dream." As sung by the production's protagonist, Don Quixote, it's the fever dream of a delusional and possibly dangerous Spanish nobleman, destined for disappointment. Something about it fit the American mood. The tune was covered by everyone who could lift a microphone in those days, and it eventually made its way to the top of the easy-listening charts.

The Gemini missions are not widely remembered. Al Shepard and John Glenn rode to glory in the early days of Mercury, and Apollo aimed for the moon, but Gemini was a crucial if often overlooked intermediate stage. It was when the agency got serious about working in space, when the cowboys had to start using calculators and testosterone ceded ground to technique. But as important as the accomplishments of Gemini proved to be, NASA was just warming up for the main event. John F. Kennedy's vow to send a man to the moon and bring him safely back home before the end of the decade was beginning to look less like that ridiculous *song* everyone was singing and more like an actual deadline. With Mercury and Gemini concluded, NASA turned its attention to Apollo. The agency was rapidly expanding and, in the process, soaking up as much money as Congress could throw at it. And with an ambitious schedule of Apollo flights, each of which would require a crew of three, NASA put out a call.

America needed more astronauts.

9

The Rise and Fall
of the American Astronaut

*America's astronaut corps has evolved from an elite cadre of cockpit cowboys
to a more inclusive but still highly skilled group of fliers, scientists, doctors,
and engineers. But does anyone know their names?*

America's Original Seven astronauts were introduced to the world on 9 April 1959 in a sudden supernova of flashbulbs, questions, and frantic cameramen. It's worth learning who these instant heroes were. After all, your grandparents did. They looked up from their Major Mike Meteor decoder rings and neighborhood paper route maps and marveled at the men chosen to counter the Soviet menace by journeying deep into the sea of night, where soon we'd be surveying the lush jungles of Venus and fighting the communists with laser cannons.

Joining Alan Shepard and John Glenn at the astronauts' table that day were Scott Carpenter, Gordon Cooper, Gus Grissom, Wally Schirra, and Deke Slayton. The Original Seven was an eager collection of gleaming teeth and prominent brows, tested by combat but comfortable in a cardigan, certified as competent in every way. Several sported military-style crewcuts, and their bristly hair stood like tiny ramparts situated on the high ground of their heads. Possibility flashed from their faces like lightning reflected on a northern lake. Their smiles were omnivorous. They were the nation's first superhero unit, fearless and focused, a platoon of paladins recalling Kurosawa's *Seven Samurai* and preceding (and perhaps inspiring) both the seven-member Justice League of America and Marvel's mighty Avengers. Fond of comic books? Think of Al Shepard as the Flash, or possibly Quicksilver, a slender, mercurial figure who was first to make it into space. Glenn was Captain America, spotless, forthright, and true. Schirra was a perpetual weisenheimer whose undeniable talents were sometimes outpaced by his mouth. We'll call him Spiderman. Carpenter, a cerebral type, was the Vision, Slayton Nick Fury,

Gordon Cooper Bucky Barnes, and Gus Grissom the Batman, for his intensity, engineering prowess, and occasional bursts of bitterness. The number seven has strong biblical associations, symbolizing perfection and completeness. It's also, as everyone knows, exceedingly lucky. So was it happenstance that the group totaled seven?

Please.

The New Nine joined the astronaut corps in 1962, just in time to start training for the two-man Gemini missions. Why nine? No definitive answer exists in the literature, but surely the number echoes medieval chivalry's Nine Worthies, the most excellent and exemplary of all heroes. Some say this was the most talented of all the classes, featuring the über-astronaut, Neil Armstrong; longevity's poster child, John Young; and the ever-approachable Jim Lovell. Groups three and four joined in 1963 and 1965, respectively. Group four was known as "the Scientists," as it was made up of men recruited for their intellectual accomplishments rather than their piloting skills. NASA has chosen twenty-two astronaut groups since the Original Seven were announced. At one point during the Space Shuttle era, the corps swelled to 149, but it has steadily declined since then. Each new group of would-be spacefarers has tried to create an identity for itself while carefully avoiding claims to the luster of the Original Seven or the New Nine. The names are predictably ingratiating. No "Guardians of the Galaxy" here. No "Love Lords of Jupiter" or "Moondust Junkies" have entered the program. Disappointingly, America's new astronaut candidates have adopted self-flagellating boot camp nicknames like the Maggots, the Chumps, and the TFNG, variously reported to stand for the "thirty-five new guys" or "the fucking new guys," or possibly both.

The first few astronaut groups were carefully managed. Their images were scrubbed and sanitized not only by NASA but by journalists who joined in the hero worship—some of whom, indeed, paid handsomely for the privilege. The rocket jocks did their part. They were a famously private and protective group, skilled at saying nothing, who seemed to pride themselves on presenting similar and imperturbable faces. Yet these faces were studied intently. Wally, Gordo, and company were courted and fawned over. They were invited to Rotary dinners, civic celebrations, gymnasium openings, and college commencements. John and Annie Glenn showed up in magazines and newspapers and on TV. They water-skied with the Kennedys and waved from motorcades. One rookie astro visited Pocatello, Idaho, in March of 1968 and was awarded the Indian name Chief Ride 'Um Rockets by local boosters. *Apollo 17* astronaut Ron Evans appeared with teen idol Bobby Sherman at a concert in Houston. The space men attended Houston's Spanish Ball with His Excellency Marques Mercy del Valle, Spanish Ambassador to the United States, and hobnobbed with John Wayne at the American Cancer Society's 1970 Salute to the Astronauts in Las Vegas.

It wasn't just the astronauts themselves who were scrutinized. Their wives and kids moved under the microscope as well. In this peculiar period of national insecurity, when the Russians seemed to know a lot more than we did, the astronaut family became a symbol of national identity. Americans had occasionally looked to the broods of other types of celebrity for instruction and edification, with mixed results. The Kennedys were beautiful, but their wealth was difficult for most Americans to identify with. And politicians were mediocre representatives at best—skirt-chasing publicity hogs for the most part, too eager and calculating for perfect candor. Athletes were attractive but unreliable, muscled-up meatheads with little to say—and if they played for the wrong team, who wanted to read about them anyway? Actors were worse: pill-popping bundles of insecurity with odd cravings and alarming sexual proclivities. So the astronauts were up, and the glare was blinding. This small group of apparently earnest, death-defying patriots didn't say much of interest, but it wasn't because they couldn't. Deke Slayton's reticence could be taken as modesty, Gus Grissom's midwestern minimalism for humility. Their wives—military wives, after all—were uncomplaining and dutiful, their kids goofy and adoring.

For a time the American astronauts were paragons, the nation's hometown heroes, bashful at heart but willing to fight for liberty, Chrysler, and a piece of the moon. And their wives, who'd never signed up for this, after all, kept smiling alongside them, even as their husbands' long hours, perpetual peril, and temptations to infidelity began to cause serious strains in their marriages and psyches. Most of those early astro-marriages eventually failed. But don't blame that on the wives. There is a memorial grove at Johnson Space Center for astronauts who have passed on. It's a lovely place, an appropriate reminder of mortality and shared purpose. But somewhere on this leafy campus there should also be a monument to the spouses who supported the spacefarers: a single steel girder, perhaps, warped in the middle. The warp would represent *stress*, of course—and the girder bent but never broken.

Fringe Benefits

The spacefarers were all government employees, so they didn't make much of a salary, but the perks of joining the astronaut corps made the low pay, long hours, and lurking potential for a disintegrative death worthwhile. A publishing deal signed by the Original Seven paid dividends for years afterward. The astronauts were able to lease new Chevrolets—Corvettes, for example—for a dollar a year. There were Oilers games to go to, real estate opportunities dangled by mail, oil company investment offers, junkets and freebies and low-cost loans available. Not everyone could resist sharing the details of their good fortune. Astronaut Tom Stafford, Deke Slayton's right-hand man, wrote a memo to his colleagues in October of 1969 asking them please to keep their traps shut about . . . *you know* . . . the goddamned *Corvettes* and things. "The result of [our] flight crew performance," he wrote, "has created a status where we are frequently invited to both social and professional sports functions in the Houston and surrounding areas. In addition to these functions, there are other types of intangible benefits which are afforded the group. . . . Deke and I both consider the discussion of these items outside the group to be detrimental to our overall office position in both the professional and social areas."

Intangible benefits! There was nothing intangible about the Corvette Sting Ray's 350 cubic-inch Chevy engine with 435 horses under the hood, available to astronauts with the T-top, double-barreled Holley carb, and four-wheel disc brakes for the annual cost of a cup of coffee, perfect for drag racing on the Gulf Freeway or cruising for Cape cookies on the streets of Cocoa. That was pretty damned tangible! Stafford was afraid of arousing jealousy in outsiders. It was a valid concern but probably overstated. Americans expected some glamor from their cosmic flyboys. In that brief shining period when everything seemed to be going right for the nation, astronauts were the republic's new royalty. In a photograph taken by student journalist David Chudwin during the Apollo era, astronauts Al Bean, Jim Irwin, Charlie Duke, and Bruce McCandless II stand outside a Florida Ramada Inn looking like impromptu propaganda. They're all lean and loose limbed, vaguely vulpine, an advertisement for adrenaline and aviator glasses. The astronauts were Establishment eye candy, sharp-edged and untouchable. Despite the fact that they were among the nation's best pilots, and several were combat veterans, not one was required or even allowed to go to Vietnam to kill indigenous communists—a fact that occasionally caused some uneasiness among the astronauts themselves. *Gemini 9* spacefarer Gene Cernan even *asked* to go but was summarily turned down. He was too valuable where he was. The astronauts were the president's pen pals, Tom Swift's bachelor uncles, one-half Boy Scout and two-thirds bad ass. They knew Science but didn't bring her to parties. They could fly anything. They cut their hair short and at precise geometric angles to minimize drag. It was easier to enter the future that way.

The idea of the corps as a collection of interchangeable idols, identically competent and clean cut, started to fray as the number of rocket jockeys increased. Astronaut Brian O'Leary made the point clear after he resigned from the program in 1968. A member of the group six scientist-astronaut class, O'Leary had an axe to grind. Shortly after joining NASA, he and his comrades were flat-out told by head astronaut Deke Slayton that the program would have no use for them in the foreseeable future. It was an admission that understandably dampened the class's enthusiasm.

But O'Leary was an odd choice for the astronaut corps from the beginning. He'd earned his PhD from Berkeley in 1967 with a dissertation about possible properties of the surface of Mars. Even at the time of his selection, he was disenchanted with the American military's role in Vietnam. He hated Houston, and once inducted, he complained bitterly about what he saw as the anti-intellectual prejudice in the manned space program, describing the astronaut corps as "fifty clean-cut, erect, alert" military types with little scientific inclination or imagination—unlike *him*, for example, who was much slouchier and more interesting. He wrote that forcing astronomers and physicians to fly jet airplanes, as Deke Slayton insisted they do, was a waste of time—and potentially of lives. He wasn't at NASA to fly jets. He was there to do science.

That sort of statement couldn't have sat well with Slayton. Donald K. "Deke" Slayton was born to a farm family in upstate Wisconsin in 1924, not long before Robert Goddard tested his first liquid-fueled rocket. He grew up in grinding poverty during the Great Depression and was a restless child; his mother would sometimes tie him to a tree to keep him from wandering off. A farming accident at the age of five sliced the ring finger off his left hand but never slowed him down. He enlisted in the U.S. Army Air Force during the Second World War and wound up flying B-25 bombers over Italy, logging a total of fifty-six sorties, some under enemy fire, facing death and dismemberment on a more or less daily basis.

He eventually became a test pilot at Edwards Air Force Base and was selected in the first group of astronauts in 1959. Rangy and wolflike, as gruff as something that lived under bridges, Slayton was a taciturn type who kept his words to a minimum. He was said to be so tough that he chewed on nails and spat out thumbtacks. A heart problem—he had one, mind you; it just didn't work right—got him grounded in 1962, so he missed his chance at a Mercury flight. This didn't slow him down either. His colleagues thought so much of him that they suggested he become head of the astronaut office—or "chief astronaut," which was a position that didn't exist until they invented it. Slayton eventually became responsible for picking flight crews, and he did so all through the Apollo missions. Cleared for flight in 1972, he was assigned to the Apollo-Soyuz Test Project in 1974 and finally made it to space the following year. It was Slayton who ensured that NASA's flight crews in the sixties and early seventies consisted mostly of test pilots like himself—veterans of bad breaks and freakish weather, unproven technology and extreme risk. He made no bones about it. He preferred pilots—the tougher the better. He commanded the astronaut office by force of personality. "He was like a god to us, almost," said one astronaut. And this god preferred men who, like him, could fly.

In fairness to O'Leary, four astronauts *had* been killed in T-38 accidents during the previous three years. Further, Yuri Gagarin, the first human being in space, also died in a jet airplane crash in 1968. O'Leary may have been right about the pointlessness of a piloting requirement for men who were primarily interested in examining lunar dirt, but he misjudged the individuality of his comrades. Though they were demographically similar, the astronauts of those early classes were strong-willed and disparate personalities. Within a few years, *Apollo 9* spacefarer Rusty Schweickart would be studying transcendental meditation. By 1979, as a science advisor to California governor Jerry Brown, he was "sitting in a Pasadena auditorium with a metallic star pasted on his forehead as dancers circled him, chanting for the elimination of nuclear power plants." Later still, he became a sort of cosmic Cassandra, a leading voice in warning the world about the need for technology to identify and deflect large asteroids headed toward Earth—a warning that NASA is at last taking quite seriously.

Apollo 7 astronaut Donn Eisele took a position with the Peace Corps after leaving NASA. Mike Collins departed Houston to join the U.S. State Department and subsequently ran the Smithsonian's new Air and Space Museum.

Jim Irwin became an evangelist, and later led several expeditions up the slopes of Mt. Ararat in Turkey, searching for the remains of Noah's Ark. Al Bean became a painter who sprinkled moon dust into his creations. Ed Mitchell conducted experiments in extrasensory perception (ESP) while in space. He had an epiphany on his return from the moon on *Apollo 14*, and later established the Institute of Noetic Sciences to explore the ability of some individuals, as one institute newsletter put it, to "perceive information not presented to any known sense and blocked from ordinary perception." Despite what O'Leary described, the astronauts weren't really automatons, devoid of imagination or eccentricity. They just had to act like they were.

There were good reasons for the conformity. From the earliest days of NASA, seat assignments on America's rockets were hard to get and difficult to predict. Before Al Shepard's flight in 1961, the agency famously designated a three-man subgroup of the Original Seven, announcing that one of the trio would become the first astronaut in space—but declining to say who this would be until as close to launch day as possible. Astronaut Walt Cunningham discussed the crew-selection guessing game at length in his 1977 book *The All-American Boys*, while conceding in the end that he was never quite sure how the process worked—even when, early on, it favored him with an assignment to *Apollo 7*. In his book *Riding Rockets*, three-time rocket rider Mike Mullane writes caustically about the utter inscrutability of the flight crew selection process during the early Space Shuttle years. Astronaut Kathy Sullivan, the first American woman to walk in space, has stated that "the reasoning behind the particular technical assignment each of us got [when starting at NASA] was a complete mystery to us, as would be the logic behind every other assignment in our astronaut careers, along with the process by which the decision was made." It's a theme that continues to the present day, with the candidates for NASA's planned Artemis moon missions left pondering their chances for a lunar landing as the agency secretly consults its feathers and fish bones to determine the lucky travelers.

It was a grueling existence. It wasn't just the competition that made life in the Manned Spacecraft Center's Building 4 so tough. It was the fact that the competition was so *good* and the yardsticks so unclear. In the early days of crewed space missions, every astronaut was qualified for every flight. It was like being on a football team consisting entirely of quarterbacks. Because only one (Mercury), two (Gemini), or three (Apollo) astronauts could actually go up on any given mission, competition for a flight was fierce. It wasn't just what you could do that mattered. It was also important to know what you *couldn't* do.

A number of factors could knock an astronaut out of line for liftoff. It didn't take something lurid and outlandish, like one astro's alleged diaper-clad drive from Houston to Florida in 2007 to kidnap her lover's new mistress, to lose your place in the lineup. Slayton and Al Shepard were both temporarily disqualified from mission assignments due to health conditions. John Bull, a group-five astronaut, resigned from the program after being diagnosed with pulmonary disease. Ken Mattingly had to relinquish his place on *Apollo 13* because he'd been exposed—just *exposed*—to German measles. Duane Graveline was "allowed" to resign from the program when his wife threatened to file for divorce, and *Apollo 7*'s Donn Eisele never flew again after he became the first astronaut to *seek* a divorce (though others who were close behind him were not similarly penalized). Scott Carpenter was reportedly blackballed for attempting to be both a scientist and a pilot during his Mercury mission, and Dave Scott, Al Worden, and Jim Irwin were kicked out of line after the news of the *Apollo 15* postage stamp scandal—which involved carrying unauthorized postal covers in the spaceship, allegedly for later sale—broke. Though the scandal is little remembered today, it was a big black eye for NASA in the early seventies.

It's a measure of how much things have changed since the sixties that allegations of financial wrongdoing—the "first crime in space," as journalists breathlessly called it—brought against astronaut Anne McClain in 2019 by her soon-to-be ex-wife apparently led to no negative repercussions for McClain's career. Federal investigators found the charges to be false, which helped, of course. But in the not-so-distant past, the existence of a lesbian relationship *or* a pending divorce might itself have been enough to wreck a career.

The Astronaut Whisperers

Deke Slayton was at least being frank when he said he preferred test-pilot experience in assembling his teams. He clearly had his favorites, chief among them his bear-hunting buddy, big-knuckled Gus Grissom. After Slayton handed over the business of astronaut anointment in 1974, though, it was hard to tell *what* went into the flight-selection process, what lent a person that secret, sacred aura of righteous stuff that qualified him or her for a taxpayer-funded trip through the clouds. Given the quality of the competition, the relative scarcity of spaceflight opportunities, and the mysterious, seemingly arbitrary nature of crew selections, it's no wonder that active-duty astronauts attempted to blend in rather than stand out.

Slayton's successor as crew selector was George Abbey, who was an enigmatic figure even to those he worked with. A 1954 Naval Academy grad, he became an officer in the air force and flew over four thousand hours in various aircraft. After working on several air force projects, including the experimental X-20 Dyna-Soar space plane, he joined NASA in 1967 and steadily worked his way up through the bureaucratic ranks. As a young man, he'd loved reading the adventures of Buck Rogers in the comics. He wanted to be an astronaut himself but was derailed by an air force requirement that he attend test pilot school first. Just as well. Abbey's greatest talent seems to have been absorbing knowledge and then knowing how and when to use it. He was a sponge. He knew everyone, and what they did. He listened intently but rarely spoke. His rise to the top at NASA led him to the office of director of flight operations in 1976. In this post he became Slayton's successor and, in some ways, his managerial opposite. Slayton was known for terse delivery and simple truths. He called it like he saw it. Abbey was by contrast congenitally secretive, someone who rarely said what he could have someone else say for him. Thick through the middle as he matured, he had a long face, hooded eyes, and a somber expression, like someone who's used to hearing bad news—or delivering it. Abbey's crew selection techniques were notoriously obscure and seemed to some quite arbitrary. He was a polarizing personality. Many astronauts *did* like him, and some received plum assignments. Others complained about him and his mandarin-like ways, his alleged favoritism, his apparent preference for naval aviators over air force pilots. He was the quintessential bureaucrat, they said. High-fives and happy hours. Not a bone in his body. But he must have been doing something right. Abbey served in NASA's upper echelons for thirty-four years, the last five as director of Johnson Space Center. He was instrumental in getting women and minority astronauts into space. And his crews, like Slayton's, were always top-notch.

Swatting the WASPS

Though it seemed natural to many Americans at the time, there were demographic peculiarities to the astronaut corps that leap out at us now. First, the astronauts were all male. In fact, a handy mnemonic for remembering the sequence of early American space programs is "M-G-A," for both *Mercury-Gemini-Apollo* and *More Guys Allowed*, as each program allowed for one additional dude to climb aboard. The astronauts were also white. Very white: McDivitts and Glenns, Armstrongs and Cunninghams, descendants of the Scots and English and Germans of northern Europe. The simplest explanation for this imbalance, then and now, is that the astronauts were drawn primarily from the ranks of test and fighter pilots, who were—yes—*also* white, male, and possessed of that ostentatious earnestness characteristic of WASP culture.

But this wasn't the only explanation. There were currents of racism and sexism that kept pilots like air force captain Ed Dwight, female aviator Wally Funk, and who knows how many other aspiring spacefarers from getting a shot at the Big Beyond. Numerous tantalizing *What Ifs* bubble up to us from those years. Some of them form fizzy plotlines in the television series *For All Mankind*. Others are a bit more poignant. For example, Major Robert Lawrence, an accomplished pilot and the holder of a PhD in chemistry from the Ohio State University, might well have become America's first African American in space. He was selected to be an astronaut in the air force's Manned Orbiting Laboratory program in June of 1967, but MOL never flew. The project lost funding and seven of its astronauts—all of whom went on to fly on the space shuttle—were transferred over to NASA in 1969. Lawrence might well have been transferred as well. Unfortunately, he died in a plane crash during an instructional flight in December of 1967.

Astros Reconsidered

The image of the rock-jawed, steely-eyed astronaut took a hit in the late sixties and seventies. In popular music, fiction, and films, he was no longer unquestionably virtuous and invincible. Far from it. Popular imaginings became a little like Greek tragedy, with the hubris of our spacefarers, real or imagined, subjecting them to all manner of bizarre and undignified threats. Thus, an astronaut could be manhandled by glorified gorillas, as Charlton Heston was in 1968's *Planet of the Apes* (and he was the *lucky* one—one of his crewmates was killed by primates, stuffed, and displayed in an ape museum). He could regret having to go to Mars, which, it should be noted, is cold as hell. He could suffer a mental breakdown, as in the 1969 movie *Marooned* and Gary Malzberg's 1971 novel *The Falling Astronauts*, be killed by a computer, as in Stanley Kubrick's 1968 tour de force *2001: A Space Odyssey*, brutalized by bikers, as in the 1978 novel *The Ninth Configuration*, or extorted and murdered by NASA, as in 1977's box-office smash *Capricorn One*. Then, too, the astronauts played a role in their own demystification. They wrote about themselves. Most notably, Buzz Aldrin opened up about his post-Apollo depression and alcoholism in the 1973 memoir *Return to Earth*. They also wrote about each other, and not always in flattering terms. The astronauts, it turned out, were people. They had problems. Not necessarily *being-captured-by-apes* problems—by and large, the spacefarers went on to live productive and sober post-flight lives. But the sheen faded. Just as in the culture at large, white no longer meant spotless, and men no longer seemed to deserve a monopoly on important opportunities.

Still, it wasn't until 1978 that the agency took a significant step to diversify the astronaut corps. Drawing in part on the efforts of Nichelle Nichols, the African American Lieutenant Nyota Uhura on the television series *Star Trek*, the agency encouraged women and people of color to apply for a spot in the space program. Lo and behold, the thirty-five individuals selected to be astronauts in group eight in 1978 did indeed include women, African Americans, an Asian American, two Jewish Americans, and even a gay American, Sally Ride—though Ride's sexuality remained a carefully tended secret until shortly after her death. In short, it included all manner of Americans, and was the start of a much more purposeful effort at inclusion that continues today.

The addition of women to the astronaut corps was especially impactful. The Soviets famously sent the first woman into space in 1963, and it was a Russian who made the first female space walk in 1984. But the Russian program has otherwise been so male dominated that it's hard to see the feats of Valentina Tereshkova and Svetlana Savitskaya as anything other than stunts. The United States has been considerably more eager to employ women in meaningful space-related roles. Fewer than a hundred women have traveled into space, including both suborbital and Earth orbit flights. (At present, no women have traveled beyond Earth's orbit, but the Artemis program promises to change that.) That's not a happy number. While only approximately 14 percent of space travelers worldwide are women, the majority of female astronauts hail from the United States. Furthermore, American women have commanded shuttle missions, set space endurance records, and run the Johnson Space Center. This is not inclusion for inclusion's sake. It's integration—and, indeed, *leadership*.

It bears noting that while the U.S. space agency has become more inclusive, NASA still hasn't progressed in its views of the eligibility for astronaut training of individuals with physical differences. Naval aviator and double amputee Frank Ellis might have made a good astronaut but wasn't seriously considered when he applied for inclusion in astronaut group five in 1966. The European Space Agency recently selected a Briton named John McFall, who lost a leg in a motorcycle accident, as an astronaut. Perhaps NASA will expand its views on differently abled astronauts as well.

Aside from occasional firsts—Scott Kelly's yearlong sojourn in space, for example, or Chris Hadfield's acoustic Bowie cover on the International Space Station—the exploits of American and other astronauts are mostly anonymous these days. As the journalist Bryan Burrough wrote in 1998, "Today, swathed in the smothering layers of NASA's safety bureaucracy, shuttle flights pack all the suspense of a crosstown bus. They are routine. No one other than science teachers, *Star Trek* fans, and documentary filmmakers much cares what the astronauts do in space. 'Looking at stars, pissing in jars,' is the snide catchphrase for astronaut work you hear at Kennedy Space Center."

Ouch.

Burrough's assessment was made at a time when there was a much larger astronaut corps than today, and shuttle missions had become commonplace. Today, by contrast, there's a building buzz about America's return to the lunar surface, and who will get there first. There's even renewed interest in the fact that almost no one really knows how that determination will be made— including the astronauts involved. Buzz is good, in NASA's eyes, as it increases public attention. Smiling spacefarers help the agency obtain funding for its less glamorous endeavors, like asteroid probes. But the adulation heaped on the Original Seven is a thing of the past. And that's okay. There's a bright side to the fading glamor.

The astronauts were never as important as we wanted them to be, and our fascination with the men and women at the top of the rocket tends to obscure the engineers and analysts, flight controllers, mechanics, and myriad other individuals who make spaceflight possible. Now when we see our astronauts preparing for launch or floating around in the microgravity of the International Space Station, they look not like demigods but like our neighbors and cousins and friends. In fact, some of them *are*—or may soon be. As private companies like Virgin Galactic, Blue Origin, and Axiom offer tourist flights into space, the distinction between regular people and astronauts is blurring. And maybe that's the point. NASA's current crop of astronauts isn't entirely like us. They're enormously intelligent and talented, and they're fanatically dedicated to their jobs. Not all of them fly jet aircraft. Some—well, *one*, anyway—studies unicellular life forms that can take on the consistency of gelatinous slime. *Slime molds*, as they are understandably referred to, used to be considered fungi, but they are now categorized as being part of a number of biological supergroups, sort of like a unicellular Eric Clapton. They might live beneath the surface of our solar system's icy moons. The point is, our astronauts look like us, and they talk like us, and that's enough to get anyone, male or female, Anglo or Asian or African American, dreaming about what they can accomplish in space. This seems right.

After all, not every hero has a crew cut.

The Best Space Stuff You Can Watch

1. *Apollo 11*: There's no substitute for the real thing. This 2019 documentary directed by Todd Douglas Miller proceeds without voice-over narration or talking heads, using contemporary film footage to chronicle humankind's greatest adventure in beautiful and breathtaking detail. Covering a lot of the same ground as *Apollo 11* but weighted with a determined effort to grok the meaning of the moon landings, Theo Kamecke's 1972 documentary *Moonwalk One* is also worth a look. Some of the film footage is so lush, you'll feel like you wandered into an issue of *National Geographic*.

2. *Apollo 13*: Ron Howard's gripping, more or less accurate account of the "successful disaster" that almost took the lives of three American astronauts serves up lots of nerve-jangling nostalgia as we watch NASA's engineer army work out plans to save Lovell, Haise, and Swigert from the world's first in-space disaster.

3. *Good Night, Oppy*: The most emotional robot movie this side of *WALL-E*, this 2022 documentary, directed by Ryan White, chronicles the "lives" of two Mars rovers. Relentlessly manipulative, right down to its muscular pop music soundtrack, *Good Night, Oppy* will make you forget all those bad things you thought about the coming singularity. Well, most of the bad things.

4. *Star Trek*: Cheesy as fondue night at the local Elks Lodge, over-acted and under-produced, the short-lived sixties TV series is nevertheless required viewing for any American space nut. Come for the groovy aliens and Mod Squad–era miniskirts but stay for the multiethnic crew and relentless optimism about humanity's future in space. Once you get started, you'll want to check out the myriad *Star Trek* succes-

sor series, some good, some not so good, the many *Star Trek* movies, and the magnificent cinematic parody, *Galaxy Quest*. Gene Roddenberry's ungainly brainchild has truly lived long and prospered. Also worth watching on the small screen: *For All Mankind*. This revisionist soap opera series takes space history and gives it a little shake. Well, maybe a *big* shake. The Soviets get to the moon before we do, the North Koreans hot foot everyone on the next leg of the journey, and the future is altered in all sorts of interesting and melodramatic ways, both serious and silly.

5. *The Right Stuff*: Based on the much better book by Tom Wolfe, NASA insiders mostly hate director Philip Kaufman's cartoonish account of the Mercury program and the blue-eyed boys who rode the rockets. John Glenn reportedly called it "Laurel and Hardy Go to Space." The rest of us have made it a favorite.

6. *2001: A Space Odyssey*: Despite the famously opaque ending, there's no disputing the gorgeously detailed imagining of a von Braunian future in Stanley Kubrick's 1968 film. And HAL is still the most menacing robotic villain in the movies not portrayed by Arnold Schwarzenegger.

7. *Searching for Skylab*: Like *Skylab* itself, Australian filmmaker Dwight Steven-Boniecki's low-budget but loving look at America's forgotten space station, our microbus in the sky, deserves more attention.

8. *Destination Moon*: We've come a long way, baby! Science fiction set in the future can leave us frustrated with our lack of technological progress. This film has the opposite effect. Directed in 1950 by Irving Pichel with input from sci-fi scion Robert Heinlein, *Destination Moon* was an attempt to portray a moon landing more or less realistically. Compare it with the real thing, which took place less than twenty years later, and you'll see that we *are* making progress. If you feel like going even further back, check out *Frau im Mond*, Fritz Lang's 1928 depiction of a co-ed flight to the moon. (And yes, there's a love triangle.) It's a fascinating film, full of portents of a spacefaring future, but it moves so slowly that you might want either to speed up your viewer or slow down your brain.

9. *Mission Control: The Unsung Heroes of Apollo*: A 2017 documentary based on Rick Houston's book *Go, Flight!*, *Mission Control* relives the exploits of the engineers and technicians who controlled America's early space missions, from Mercury through Apollo. Through commentary by the individuals themselves, we learn some of the secret holy lore of the Houston trenches, from the import of computer alarms 1201 and 1202 to the significance of *SCE to AUX* and "Houston, we've had a problem." Director David Fairhead serves up a fascinating look at the tough, creative, nicotine-driven individuals who saved NASA's bacon on more than one occasion.

10. *The Martian*: Far-out and freaky as it often is, most sci-fi has only a passing familiarity with the nuts, bolts, and turbopumps of actual space travel. Adapted from Andy Weir's imaginative if occasionally sophomoric novel, director Ridley Scott's *The Martian* does a nice job of portraying the various complications that will be involved in a mission to the Red Planet, trying hard to get the facts straight, even as it pushes them in the service of a great story. This one's got jetpacks, radioisotope thermoelectric generators, and way too many Martian potatoes.

11. *Contact*: Jodie Foster stars in this adaptation of the novel by space guru Carl Sagan, which explores how humanity might be contacted by an extraterrestrial civilization. Matthew McConaughey violates everyone's personal space as surf theologian Palmer Joss, who loves Foster's character so much that he (*Wait, what?*) sabotages her fondest dream. Director Robert Zemeckis's film eventually detours off into Kubrick-land, but in the meantime, it's clever, gripping, and fun. In the same vein, the 2016 Denis Villeneuve film *Arrival* is also thought-provoking and equally trippy. And don't forget *The Day the Earth Stood Still*, the 1951 classic in which a well-dressed alien lands his flying saucer in Washington DC in order to give us all a stern talking-to about humankind's aggressive tendencies—after which we, um, shoot him.

10

Apollo and the First Man on the Moon

The Apollo program remains America's iconic accomplishment in space.
This is both a blessing and a mild curse for NASA administrators,
who have been trying to equal Apollo's achievements ever since.

For a decade, the Soviet space program was the most ambitious and dynamic in the world. Sputnik was an obvious blow to the American psyche. But the USSR achieved other remarkable feats as well, not only in the earliest days of space travel but also for years afterward. For example, Moscow put the first remote-controlled mechanized rover on the moon in 1970. *Lunokhod I*, an eight-wheeled robot that looked like a cross between a baby buggy and a makeup compact, was spectacularly successful. It operated for almost a year, traveled a distance of over six miles, and sent back some twenty thousand television images of the lunar surface, including 211 high-definition panoramas.

The USSR also managed to land a probe on Venus in 1970 and, with *Venera 9* in 1975, became the first nation to obtain photographs from the surface of that disagreeable orb. *Venera 9* operated for almost an hour before the Russians lost contact with it and it was thereafter destroyed by Venus's 867-degree Fahrenheit surface temperature and massive atmospheric pressure. The photos are interesting but uninviting. Venus, it turns out, is a fixer-upper. Or, as Elon Musk has put it, "Venus is not at all like the goddess."

The race to put human beings on the moon was in no way a settled matter until late in the contest. The Soviets had their own German engineers, after all, along with a host of homegrown technicians and a brilliant rocketry man named Sergei Korolev, who survived six years of imprisonment and forced labor under Josef Stalin and went on to head the nation's space program under Khrushchev. NASA seemed to take the lead in the contest at some point during the Gemini program, when American astronauts proved they could rendezvous and dock their capsules with other spacecraft, a skill that would prove crucial for lunar missions. But ultimately, the Soviets failed in their quest to beat the United States to a crewed lunar landing—indeed, to get their cosmonauts near the moon at all—because they were unable to produce a "heavy lift" rocket powerful enough to boost a trans-lunar spacecraft and crew to Earth's satellite, a quarter-million miles away. Their best effort, the kerosene-fueled N-1 rocket, was comparable to NASA's Saturn V in many respects except the most important one: it didn't work. On 4 July 1969, just days before *Apollo 11* was launched, a test of the Russian rocket ended in a cataclysmic explosion and fire that leveled the N-1's launchpad. Though the Soviets kept testing the rocket well into the seventies, the mighty N-1 never made a successful flight. After a final failure in 1972, the program was abandoned. The Soviets had already turned their attention to other space projects. Forget lunar landings, they said, shortly after *Apollo 8* successfully journeyed to Luna and back. We've always been more interested in space stations. Few believed it. In the meantime, they left the moon with a single, star-spangled suitor.

Apollo 1

It was of course wrong to name the lunar landing program after the golden deity called Apollo. The son of Zeus was perfect—the personification of beauty, the inventor of music, a healer and protector who was eventually identified as the bringer of the sun and even as the sun itself. Some scholars say he was a model for Christ, a sort of conceptual scaffold used by early evangelists to explain certain attributes of the Savior. But why invoke the sun god for *moon* missions? One could well wonder why we name rockets—Juno, Thor, Saturn— after gods in the first place. What sort of traits are we honoring, after all? Seeing as how we *made* the rockets, are we honoring the gods, or are we honoring ourselves? Even accepting the *giant-explosive-cylinder-as-deity* convention, though, why not name the moon rocket after Apollo's pale sister, Artemis, goddess of the hunt? Perhaps NASA was hedging its bets. The moon was the goal of Apollo, but the goal wasn't a sure thing. The only certainty was that Wernher von Braun's Saturn V rocket was going to be so massive and powerful that it would drag its own sun-like fire through the sky. Sources say little about NASA's choice, the brainchild of longtime manager Abe Silverstein, other than that it seemed to fit the godlike scope and ambitions of the program. And indeed, Apollo would become famous as the Main Event, the Big Show, America's crowning technological achievement.

But it began with a disaster.

Astronauts Gus Grissom, Ed White, and Roger Chaffee spent most of the afternoon of 27 January 1967 practicing for their launch on the first crewed Apollo mission. Seated high above Launch Pad 34 at Kennedy Space Center, the three men were wearing pressure suits and were sealed in the spaceship's command module as they would be at liftoff, which was scheduled to take place just a few weeks later. The atmosphere around them was pure oxygen at higher than ambient pressure. When a short in the capsule's wiring arced, oxygen-fed flames flashed through the command module's nylon netting and foam pads. The crew tried to get out, but the capsule's hatch was designed to open inward, an action that was made impossible by the sudden rise in air pressure in the capsule as the fire intensified. The astronauts died in just a few seconds.

Gus Grissom was the second American to fly in space. Compact and grizzled, a Morlock amid the Eloi, his Mercury mission was sandwiched between the accomplishments of his more glamorous colleagues and received less fanfare. Nevertheless, Grissom was a talented pilot and engineer, popular with the program's contractors for his blunt speech and willingness to get his hands dirty. Ed White was a gifted athlete, brilliant and beautiful, perhaps the most promising of NASA's second generation of spacefarers. His space walk in 1965 vaulted him to international fame. Fresh-faced Roger Chaffee was the rookie of the crew, inexperienced but bright. NASA controlled what the public heard about the tragedy, as it would with subsequent disasters. The official line was that Grissom, White, and Chaffee perished instantly. Insiders knew this was a lie. The men screamed for help as the flames flashed through the cabin.

It was an unspeakable event. It was NASA's *Hindenburg* disaster, the agency's original sin, and its aftereffects lasted for years. In 1971 Betty Grissom, Gus's widow, did the unthinkable when she filed a $10 million product liability lawsuit against the contractor that built the *Apollo 1* command module. It was the largest product liability suit ever filed in an American court. Mrs. Grissom alleged that the contractor, North American, had produced a faulty spaceship. But there was an implication in the pleadings as well that NASA had been asleep at the wheel, unable or unwilling to correct North American's work. The lawsuit was eventually settled with a payment to Grissom of $350,000. NASA went on. But the agency's reputation for competence and, indeed, *caring*, took a significant hit. Nor was this the only repercussion. Ed

106 | APOLLO AND THE FIRST MAN ON THE MOON

White's wife, Pat, never quite got over the shock and sadness of her husband's death. After a long struggle with depression and substance abuse, she died of an overdose in September of 1983, one last victim of the launch-pad fire.

The *Apollo 1* tragedy cast a long, sobering shadow over the men who were waiting to fly. Astronaut Bruce McCandless II later admitted the obvious. The astronauts were too optimistic. In their eagerness to ride rockets, see the solar system, and beat the Soviets, they were willing to forego hard questions and to trust their lives to a loosely organized army of administrators, engineers, and contractors. McCandless was just as guilty as the rest. When he was interviewed in 1966 for a spot in the astronaut corps, he was asked if he'd be willing to travel to the moon and back with a fuel reserve of just *2 percent*. "I would," he said at the time. Of course he said it. Saying *no* was no way to get a flight. But years later, he winced at that cavalier response. Just a few months after the *Apollo 1* fire, cosmonaut Vladimir Komarov was killed when the parachutes of his *Soyuz 1* spacecraft failed to deploy properly and his capsule meteored into the earth at over two hundred miles per hour. It was a reminder, if any was needed, that nothing was guaranteed. Like Apollo's command module, the Soyuz was a new type of vessel, one that was being hurried into operation by bureaucrats anxious not to let the "other side" gain an advantage in the space race.

The dangers didn't stop the astronauts who were waiting to fly. But it did remind them of the hard fact that space travel could kill them, and that they were going to have to take a more active role in the design and operability of the machines they planned to rely on. Eventually, the pall faded. The Apollo program started up again, this time with myriad improvements, large and small, to the spacecraft. The command module was redesigned. The oxygen content of the capsule was significantly reduced while the vessel remained on Earth, and the hatch was simplified and modified to open outward so that the increased cabin pressure generated in the event of a fire wouldn't impede a rapid exit. In the meantime, those involved shared the mounting excitement of working on Apollo, the greatest engineering project in history.

Prelude

Humanity long dreamed of voyaging to visit Earth's pale sister. A trip to the moon was the archetypal expression of human adventure, and writers such as

Cyrano de Bergerac, Edgar Allen Poe, Jules Verne, and H. G. Wells all imagined the journey. De Bergerac's amateur astronaut traveled to Luna by means of a rocket-powered device, while Poe's adventurer managed to get there in a technologically advanced balloon. As we've seen, Jules Verne's spacefarers were blasted from the mouth of a giant cannon: "An appalling, unearthly report followed instantly, such as can be compared to nothing whatever known, not even to the roar of thunder, or the blast of volcanic explosions! No words can convey the slightest idea of the terrific sound!"

A hundred years later, not everyone shared Verne's enthusiasm for such man-made cataclysms. The endless summer of the midsixties ended with a notable chill. The year 1969 was marked by airplane hijackings and student protests. Richard Nixon ordered the secret bombing of Cambodia, and almost twelve thousand American servicemen died in Southeast Asia, fighting an elusive enemy in the service of an even more elusive cause. The gap between Madison Avenue's grinning and odorless America on the one hand and the realities of minority riots and the My Lai massacre on the other was becoming increasingly obvious. America was a gifted child with glittering toys, alternately idealistic and insane, high minded and hot tempered, distrustful of anyone who didn't look like it did. In Houston, though, it was hard to see beyond the bubble of adulation and energy building around Apollo, a project that combined massive amounts of human intelligence and labor with immense ambition and huge infusions of patriotic fervor.

Hundreds of thousands of Americans worked on the project. Millions more watched with fascination and growing excitement. Even the novelist Norman Mailer, exhaustingly agnostic, skeptical of the technological hubris that led to the launch, was nevertheless unsettled by Apollo's scope and audacity. Visiting the Vehicle Assembly Building (or VAB) at Cape Kennedy, where the Saturn V rockets were put together, he wrote of the awe experienced by his alter ego, Aquarius. The VAB is a Brobdingnagian structure, the brontosaurus of buildings, so big that if the doors are kept open, clouds form in the structure's upper reaches and rain falls on the workers below. For Mailer, an accomplished egotist, to admit that he was left-brain-broke by doings inside the VAB was roughly equivalent to Muhammad Ali declaring he was afraid of a fight. And yet Aquarius was humbled. As Mailer/Aquarius gushes in *Of a Fire on the Moon*, his nonfiction meditation on the *Apollo 11* landing:

The change was mightier than he had counted on. The full brawn of the rocket came over him in this cavernous womb of an immensity, this giant cathedral of a machine designed to put together another machine which would voyage through space. Yes, this emergence of a ship to travel the ether was no event he could measure by any philosophy he had been able to put together in his brain.

Earthrise

After the *Apollo 1* fire, Americans were earthbound for almost two years. A series of remote-controlled test flights paved the way for Wally Schirra, Donn Eisele, and Walt Cunningham to make the first crewed Apollo flight on *Apollo 7* in October 1968. Despite some testy interactions between the crew and mission control, the mission went well—very well. Buoyed by the performance of its machines, and acting in part on information from the Central Intelligence Agency indicating that the Soviets were getting close to a lunar landing, NASA then made the audacious, potentially ill-advised decision to send the crew of the next flight, *Apollo 8*, into orbit around the moon. It was the biggest gamble the agency ever took.

Apollo 7 had tested much of the nation's new lunar mission hardware. It all checked out—in Earth orbit, that is. What NASA asked the crew of *Apollo 8* to do wasn't Earth orbit, a hundred fifty miles above Houston. It was *lunar* orbit, 240,000 miles away, in a spaceship tested exactly once. Against long odds, Frank Borman, Jim Lovell, and Bill Anders blasted off on 21 December 1968, atop a rocket so big that Susan Borman likened the sight to watching the Empire State Building leave the planet. Borman & Co. made an almost perfect flight, orbiting the moon ten times and demonstrating conclusively that travel to Earth's satellite and back was well within the country's technological abilities. While in lunar orbit, Anders photographed Earth emerging from the shadow of the moon. *Earthrise*, as the shot is known, became an iconic image, credited by some as an inspiration for the American environmental movement. A television broadcast of each of the three astronauts reading from the Book of Genesis on Christmas Eve was also flawless—a Bible-based advertisement for American expertise and grace under pressure. It is estimated that almost a billion people saw or heard at least part of this presentation.

Two months later, in March of 1969, *Apollo 9* tested all the program's moving parts, including, for the first time, the vaguely bug-shaped lunar module, the thin-walled moon lander with four spindly legs. As this was the inaugural flight of the entire package of Apollo components, a review is in order. Astronauts Jim McDivitt, Rusty Schweickart, and Dave Scott left Earth atop a Saturn V rocket, the most powerful vehicle ever launched. It was a 363-foot-tall monster, taller than the Statue of Liberty. Weighing in at 6.5 million pounds when fully fueled, more than the combined weight of thirty-three B-52 bombers, von Braun's beast generated over seven and a half million pounds of thrust at takeoff. The roar of its launch was measured at 204 decibels, enough to kill any human being unlucky enough to be caught in the immediate vicinity. Aside from the Saturn V's three rocket stages, the spaceship it carried had three parts. One was the gumdrop-shaped command module (the CM, dubbed *Columbia* by the *Apollo 11* crew), reminiscent of the design of both the Mercury and Gemini capsules, containing a cabin for the three astronauts. This was the only part of the spacecraft that returned to Earth. There was also a service module (the SM), which supported the command module with propulsion, electrical power, oxygen, and water. Together, the command and service modules were referred to, naturally, as the command-service module, or CSM. Finally, there was the squat, ungainly lunar module (the LM, which the *Apollo 11* crew called *Eagle*), which itself had two parts—a descent stage for landing on the moon and an ascent stage to blast the astronauts back off the surface and into lunar orbit.

Apollo 9 was a phenomenally successful mission, marred only by astronaut Rusty Schweickart's nausea during the flight. And while *Apollo 9* stayed in Earth orbit, *Apollo 10* traveled to the moon for its tests. There, astronauts Gene Cernan, Tom Stafford, and John Young practiced rendezvous procedures, including a simulated landing by the lunar module, as they orbited Luna. The mission's return to Earth was clocked at almost twenty-five thousand miles per hour—still a record for the fastest human spaceflight. And now, finally, the stage was set. As technically demanding as *9* and *10* were, they were still only rehearsals for the big event. On *Apollo 11*, astronauts Neil Armstrong and Buzz Aldrin would attempt what had long been thought impossible.

At this point, having surveyed the metallic pedigree of the moon rocket and its parts, perhaps we should pause and acknowledge that all the metrics and measurements of even the most advanced space hardware pale in significance next to what is, for many people, its fundamental lure: the surging, tear-your-face-off, awesome elemental slow-motion-thunderclap power of it all. To have seen a Saturn V launch is to have known terror and delight in equal measure. The good earth shuddering. Birds fleeing en masse. An artificial sun awakening, the air crackling, the wash of sound like a wave breaking over you as you stand transfixed and affrighted by what human beings have wrought and—Lord help us—barely control. And inside the rocket the lucky ones, heroes and victims, the ones who might just die a second later, belted in for the best carnival ride ever, rising godlike up out of the swamps to dance weightless in the night. That's the space program stripped of its analytic clothing: Cherry bombs and bottle rockets, druid fires and midnight drums. The ear-stunning shriek and sudden sharp island of light weren't unfortunate side effects of a Saturn V launch. For tens of thousands of people who gathered to watch and sweat and *three-two-one* it as the moment of detonation grew near, the fire and fury were what Apollo was *about*. The power. The purity. It was Prometheus with a pipe bomb. It was a bonfire on the hilltop to drive off the night. It was an unbaffled muffler, the thrum of a Harley, the scream of a jet. Maybe this primal pushback, this denial of the darkness and cold and infinite stillness of forever, is what all rocket launches are about, why engineers create them and we continue to watch. It's simple. We enjoy it.

One Small Step

Apollo 11 was commanded by Neil Armstrong, with the tightly-wound Buzz Aldrin as lunar module pilot and cerebral Mike Collins serving as command module pilot. The most elusive and phlegmatic of Apollo's heroes, Armstrong was magician, mute, and walking monument, a man who sometimes seemed to be observing the rest of the human race through the lens of a powerful telescope, baffled by our customs. Even before he joined NASA, he'd suffered through hair-raising danger and miraculous escapes. He lost part of a wing of his F9F Panther during a Korean War combat mission and parachuted to safety. As a test pilot, he found himself miles off course when his rocket-powered x-15 aircraft "bounced" off of Earth's atmosphere on its descent from a climb above two hundred thousand feet. When the x-15 eventually fell low enough for Armstrong to regain control, the rocket plane was out of fuel, and he found himself on a flight path into heavily populated areas of Southern California. With no viable options for landing in Venice Beach or Rancho Cucamonga, the young pilot was forced to make a long, agonizing glide back to Edwards Air Force Base, where he barely reached the landing strip.

A thruster malfunction on *Gemini 8* almost killed him and his crewmate Dave Scott until Armstrong figured out, in the middle of being spun into unconsciousness, how to stop the tumble that was turning their capsule into a blender. And in May of 1968, just over a year before the launch of *Apollo 11*, he had to eject from a test flight of the lunar landing research vehicle, the so-called flying bedstead, and parachute to a landing in a scrubby field just north of the Manned Spacecraft Center. Bruised and bleeding from the mouth, the kid from Wapakoneta, Ohio, dusted himself off, changed clothes, and headed back to the office to handle some paperwork.

People admired Armstrong for his pleasant, if occasionally distracted, temperament, even as some questioned how, exactly, a normal human being could function in the face of such hazards. Mechanical misfortunes hounded him like the furies in an old Greek play. In a 1970 interview for the British television program *The Sky at Night*, host Patrick Moore peppers Armstrong with pointed questions about his space journeys and all the terrifying magic and mysteries that must have been involved. The American responds with facts, figures, and measured words as the interviewer grows increasingly manic. Moore's words come faster. The Brit assumes a sort of crouch, now, a *predatory stance*, as he tries to subdue his prey with the very weight of his wonderment. What was it like? What could you *see*? Weren't you concerned about *unsafe areas*? Unsafe areas! Death and destruction! A dusty grave a quarter-million miles from Earth!

The camera reveals that the mild-mannered astronaut has put on weight since his flight. There's a Buddha-like calm to his demeanor, a sort of fleshy equanimity, and when he occasionally searches for words, he smiles slightly when he's able to find them. He has passed through the shadow of the moon! He has seen the seas from space! And yet here he is, and he's happy to answer as best he can, and he's amused but not intimidated by Moore's aggressive eyebrows and wide-eyed inquisition. What is it about him? What makes him different? It's as if Moore wants to tear off Armstrong's head and peer inside, to gaze at the cosmic images imprinted therein, understand at last the weird expansive psychoses of space itself. And still Armstrong gazes back with the air of a marketing rep who's telling the story of his train ride to Topeka. Who IS this man? Moore thinks. You can almost hear him shouting inside. WHY IS HE LIKE THIS? It was a question many asked, and that no one has ever quite been able to answer.

The Landing

The astronauts lifted off from Kennedy's Pad 39A on 16 July 1969. After shedding the first two stages of their Saturn V, the crew began their journey to the moon (their "trans-lunar injection") via firing of the third stage of the rocket. Not long afterward, the astronauts in their CSM separated from the stage. CM pilot Mike Collins then turned the CSM around 180 degrees and went back to retrieve the LM from its position at the tip of the third stage. Once the CSM had docked with the *Eagle*, like a partygoer wearing a funny hat, the conjoined command, service, and lunar modules moved away and cruised moonward separate from the stage, which either continued to travel until it went into solar orbit or, on later missions, was crashed into the moon. The astronauts sojourned for three days in their ungainly vessel until they entered lunar orbit. The next day, Armstrong and Aldrin crawled into the *Eagle*. They said their goodbyes to Collins, separated from the CSM, and headed off toward the moon's Sea of Tranquility. This feature is not and never has been an actual "sea." Rather, it is one of a number of dark patches on the moon—the Sea of Storms, the Sea of Rains, and so on—that early astronomers assumed was a body of water.

Armstrong was at the controls, in command of the ship, standing because the LM had no seats. Program alarms were sounding. He was low on fuel, and rocks the size of carnival rides littered the glowing surface. No matter. Armstrong kept looking. He could hear the alarms. Yes, he was aware of the fuel thing. Earth was watching. *Buzz* was watching. The work of a decade and the journey of a quarter-million miles and now it looked like it might all end ingloriously in the last hundred feet. But Armstrong might as well have been shopping for a gently used Chevy Impala. It was the antifreeze again, coursing through his veins. He found a flat spot and steered toward it. He continued the descent. The *Eagle* settled into the lunar dust like it belonged there.

So it was that a short while later, on 20 July 1969, Neil Armstrong eased himself down the ladder of the *Eagle* to set foot on the moon. We take it for granted now. At the time, though, it wasn't entirely clear whether an astronaut could even *stand* on the moon. How deep did the dust go? Were volcanic lava tubes lurking beneath the gritty surface of this lunar hellscape, just waiting for some unlucky earthling to tumble in? Some 650 million people around the world were stuck to their television sets as Neil prepared to find out. It seemed to take an hour for him to get those last few feet down the ladder. NASA's transcript of moon-to-ground communications that day captures the mix of mundane and magnificent:

> 04 13 23 38: Armstrong: I'm at the foot of the ladder. The footpads are only depressed in the surface about 1 or 2 inches, although the surface appears to be very, very fine-grained, as you get close to it. It's almost like a powder. Down there, it's very fine.

> 04 13 23 43 Armstrong: I'm going to step off the ladder now.

> 04:13 24 48 Armstrong: That's one small step for man, one giant leap for mankind.

> 04 13 24 48 Armstrong: And the . . . the surface is fine and powdery. I can pick it up loosely with my toe. It does adhere in fine layers like powdered charcoal to the sole and sides of my boots. I only go in a small fraction of an inch, maybe an eighth of an inch, but I can see the footprints of my boots and the treads in the fine, sandy particles.

> 04 13 25 30 [CAPCOM]: Neil, this is Houston. We're copying.

> 04 13 25 45 Armstrong: There seems to be no difficulty in moving around as we suspected. It's even perhaps easier than the simulations at one-sixth g that we performed in the various simulations on the ground. It's actually no trouble to walk around. Okay. The descent engine did not leave a crater of any size. It has about 1-foot clearance on the ground. We're essentially on a very level place here. I can see some evidence of rays emanating from the descent engine, but a very insignificant amount.

Armstrong apparently bungled the delivery of his triumphal pronouncement. It was meant to be one small step for *a man*, not for man generally, since for man generally it was a *giant leap*. No matter. Everyone knew what he meant. It was one of the great and truly magical moments of human history, and Houston's mission controllers knew enough to keep quiet and let it unfold. Newsman Walter Cronkite chuckled. Then he cried.

The world was quiet too. Well, not *entirely* quiet. According to space historian Teasel Muir-Harmony, in Rio de Janeiro, church bells rang out through the city after news came that the *Eagle* had landed. In Thailand, prisoners released from jail refused to leave, as they wanted to stay and watch the moon-walkers on the jail's TV. Italy reported no robberies on the night of the lunar landing—presumably because everyone was inside following the adventures of Armstrong and Aldrin. The world looked on in awe and bemusement as two little figures moved stiffly across their screens like odd creatures in a black-and-white aquarium, and everywhere on the globe, watchers periodically glanced up at the sky to recalibrate the experience, to remind themselves that they weren't simply dreaming, that we'd reached out and touched Luna at last.

The first walkabout on the moon lasted two and a half hours, just a little longer than the John Wayne movie *True Grit*, which played to packed theaters across the country that summer. Neil and Buzz, as everyone now referred to the space explorers, deployed scientific instruments and gathered rock and soil samples. They tested how best to walk on the lunar surface. They took photographs and planted the American flag. Every minute was carefully scheduled. The astronauts did the best they could, given the constraints of their pressure suits and the sheer surrealistic wonder of being present in a place that had existed for millennia as a kingdom of dreams. It was hard to stay focused, even for such exquisitely precise and attentive individuals as the first men on the moon. Every step was astonishing. Each vision was a voyage of its own. The lunar surface lay before them airless and impassive, scarred and scorched by asteroid strikes, utterly silent despite the eons of impact.

The excitement only increased as the astronauts headed home. When asked how he evaluated the importance of the lunar landing, Wernher von Braun took a short leave from his extended campaign of reinvention as a born-again Christian, saying, "I think it is equal in importance to that moment in evolution when aquatic life came crawling up on the land." No less an authority than Robert A. Heinlein, the dean of American science fiction writers, called the landing "the greatest event in human history, up to this time. This is—today is New Year's Day of the Year One." Even President Nixon, no great friend of the space program, said that the mission marked "the greatest week in the history of the world since the Creation."

Coming Back Down

So where do you go after you've been so high you can touch the moon? This was the question the success of *Apollo 11* posed for NASA and indeed the United States in general. The answer, it turned out, was down.

The trajectory played out with particular resonance in the case of Buzz Aldrin, the second person to walk on the lunar surface. He never quite got over the drama of *Apollo 11* or how close he came to being first. He was just as smart as Armstrong—*smarter*, maybe. He was brilliant and driven, the first astronaut to earn a doctorate, nicknamed Dr. Rendezvous by his peers. He was strong, athletic, the best of the best. But Neil was just . . . *Neil*. Equal parts ice and intellect, calm as a cactus garden, impossible to injure or impede. Not long after returning from the Sea of Tranquility as the nation's number one hero, Armstrong retreated to academia. He carried his legend in a briefcase, took it out on ceremonial occasions, preferred to churn through engineering equations on a faded lecture hall chalkboard. But bouts of depression and alcoholism plagued Aldrin. The man who observed that the moon was a scene of "magnificent desolation" looked inside and seemed to see the same thing. He divorced his first wife, Joan, and married three times afterward. He sold cars, did commercials, dabbled in acting. He bounced from job to job. Hounded by a conspiracy-spouting moon-landing denier, Buzz drove him back with a brick-like right cross. (Okay, so this part wasn't so bad.) He danced with the stars. He hosted a professional wrestling show and got whacked with a chair. He wore suspenders, jeans, and a T-shirt that said GET YOUR ASS TO MARS. He was always charging forward, even when he seemed to be traveling in circles. Neil obtained secular sainthood. Buzz became the space community's middle child, subject to eye rolling as well as reverence, beloved but bewildering, a deity best experienced in small doses. A trip to the moon could change a man. But maybe it could also lock him in place.

Six lunar landing missions followed *Apollo 11*. Most of them conformed to what came to seem like routine: fiery liftoff, successful landing, our guys bounding around stiffly in the silvery haunted house high above our heads. Flags and footprints. Many rocks collected. Then the pixelated burst of the lunar ascent stage separating from its four-legged undercarriage like the carapace of some squat-bodied beetle and levitating into the cosmos, a sigh of relief, the long trip home, red-and-white chutes opening over the ocean as another teardrop fell from the sky.

Only one of the flights deviated substantially from the pattern. This was *Apollo 13*, and it deviated significantly. Trouble erupted fifty-six hours into the mission. An oxygen tank exploded, ripping away part of the service module's skin. Mission control personnel did some hasty math and determined that there wasn't going to be enough oxygen or electrical power in the CSM to keep the crew alive for long enough to get them back home safely. What followed was a marvel of ingenuity, teamwork, and mounting desperation. The crew moved into the lunar module. Mission control plotted a new trajectory to sling the craft around the moon to give it the impetus for its return trip, using as little electrical power as possible. When carbon dioxide levels crept up to dangerous levels inside the spacecraft, the crew, acting on directions from Houston, rigged a makeshift filter out of hoses, tube socks, and duct tape to trap the harmful gas. NASA came perilously close to losing another three astronauts. In the end, though, the crew of *Apollo 13* returned safely to Earth.

Four more moon shots came afterward. Counting *Apollo 11* and *12*, a total of twelve astronauts walked on the moon, where they collected 842 pounds of lunar rocks, core samples, pebbles, sand, and dust. Neil and Buzz remained on the lunar surface for twenty-two hours but walked on the moon for only two and half of those hours. By contrast, John Young and Charlie Duke of *Apollo 16* spent *three days* in the moon's Descartes Highlands in April 1972. Counting the time they spent traveling in the dune buggy–like lunar rover, the two astronauts logged over twenty hours outside their spaceship collecting rock samples and studying what scientists thought might be an area of past volcanic activity. While the lunar rock and soil samples brought back by the Apollo astronauts aren't visually arresting, they have yielded valuable information about the moon's age, composition, and origins, and they continue to be studied. In 2022 scientists were able to grow plants in lunar soil—an exciting development as NASA prepares to send astronauts back to the moon in the next few years to establish an extended, possibly permanent, presence there.

Fifty years later Gene Cernan remains the last person to stand in that asteroid boneyard. It was his crew—they tend to take joint credit—that took the astonishing photograph of an azure Earth called *The Blue Marble*, a sort of glamor shot of the planet awash in its oceans and wreathed in swirling clouds. Notable in part for its backdrop of black and the fact that no national borders are visible—all such demarcations are, of course, notional—it has become one of the most reproduced images in history. Before he returned to his lunar module to begin the journey home with the rest of the crew of *Apollo 17*, Cernan scratched the initials of his daughter, Tracy, in the dust. It was 13 December 1972. As there is no wind on the moon, the initials are presumably still there.

Cernan later wrote poignantly about how he was so wrapped up in getting to the lunar surface that he wasn't always a good father. He wasn't the only one. A generation of space-crazy engineers, technicians, and astronauts burned through the sixties without bothering to look around. It seemed like a metaphor for Apollo in general. Now the program was over, and people wondered what all the fuss had been about. America had done the impossible, but we'd done it repeatedly. Repetition made it look easy. Any miracle seems less momentous when it's reproduced. Meanwhile, the world lay in disarray around us, consumed by small conflicts, increasingly polluted, riven by racial and ethnic feuds. We were sick of our politics. We were tired of our own obsessions. In Matthew D. Tribbe's 2014 book *No Requiem for the Space Age*, he writes that citizens of the United States grew bored with the space program even as people in other nations began to appreciate it. Only a year after Neil Armstrong took his small step, for example, only one in fifteen residents of St. Louis, or 7 percent of the population sampled, remembered his name. New Yorkers were a bit more appreciative: eight in twenty-two, or 36 percent, could still identify the first man on the moon. In 1971, says Tribbe, the World Almanac dropped Armstrong's name from its index. "Whatever happened to Neil Whosis?" asked the *Chicago Tribune* in 1974.

Apollo was ambition, daring, and discipline, an extended exercise in national will and technical excellence. The glow of the program's achievements had faded long before Gene Cernan and his crew returned to Earth. In fact, it would be years before the nation learned to appreciate the moon landings again—not as a race, or to stake a territorial claim, or for military advantage, but as an expression of the human spirit faced with a seemingly unreachable goal. And ironically, by venturing outward, we gained an increased appreciation for the lonely cosmic oasis we call home. As the astronomer Carl Sagan put it, "Whatever the reason we first mustered the Apollo program, however mired it was in Cold War nationalism and the instruments of death, the inescapable recognition of the unity and the fragility of the Earth is its clear and luminous dividend, the unexpected final gift of Apollo."

11

Skylab and the Renaissance of American Science

America's first orbital station was an attempt to transition from space-as-obstacle to space-as-habitat. The craft was bruised and battered by its 1973 launch and only resuscitated through heroic actions by its crew. But Skylab *survived to deliver a surprising win for science, sincerity, and general funkiness. It was a triumph of the human spirit, which just happened to be wearing a really weird mustache at the time.*

NASA had plenty of big ideas for the post-Apollo era. Neil Armstrong predicted that we would see scientific research stations on the moon within his lifetime. One agency study conducted in 1969 laid out plans for a landing on Mars in the 1980s. Associate administrator George Mueller spoke confidently that same year about the construction of not one but *two* large space stations, one in orbit around Earth, the other circling the moon.

Alas, it was not to be. Even as the infrastructure and hardware of the Apollo program reached flight readiness, NASA's budget began to shrink, dwindling from $5.9 billion in 1966 to $3.4 billion in 1971. President Nixon was little help. Indeed, if the American space program has ever had a nemesis—a Thanos, a Voldemort—it might well be the Machiavellian president from California, who basked in the reflected glow of the Sea of Tranquility one moment and slammed the door on a host of space programs the next; it was Nixon who, for example, scuttled plans for an additional three Apollo missions. The one front-burner project that survived this period of budgetary bloodletting was the space station. And this wasn't the "hundred-man" station envisioned by NASA, Stanley Kubrick, and a shadow nation of sci-fi nerds. Like many space vessels, the finished product was a mere shell of mightier dreams. It was, in short, *Skylab*.

Officially designated as such in 1970, using a moniker suggested by air force officer Donald Steelman, the program was makeshift from the start, a sort of patchwork project constructed of clever notions and surplus parts. Science fiction writer and journalist Ben Bova called it "a jury-rigged vehicle cobbled together from the leftover pieces of the murdered Apollo program." It was criticized by some as simply a way for NASA to keep money coming in the door as the agency worked to develop its *real* ambitions—a space plane, huge orbital stations, and missions to Mars. Nevertheless, *Skylab* involved real science, real engineering, and, in the end, real fortitude on the part of its inhabitants.

Defining Skylab

If you're unfamiliar with the story of America's first space station, you're not alone. The project has always struggled for respect.

The first question most people have is simple. What *was* Skylab? Was it a thing? A program? A mission? Actually, it was all three. Maybe the simplest place to start is with the *thing*. You've probably seen photographs of *Skylab*. It looked like a flying windmill with wings—or, rather, *a* wing. This was America's first space station, a metal cylinder one hundred feet long and twenty-two feet in diameter. It was originally two separate pieces of hardware: the "orbital workshop," which was basically the long cylindrical piece; and the so-called Apollo Telescope Mount (ATM), a bolted-on space observatory that sported four solar-power-generating arrays. Perceptive viewers will also notice in most photographs that docked to one end of the station is the Apollo command and service module, which is how astronauts flew to and from their orbital home.

The Skylab program that was created to launch and operate this free-falling science platform survived the Nixon-era budget cuts for several reasons. It was cheap, since it used a lot of preexisting hardware. It was also a response to the Soviets' space station program, which put up its first orbital living quarters in 1971. And NASA was still dealing with a disgruntled constituency—scientists, who were vocally disappointed that so much money had been spent on Apollo with so little apparent attention to serious research. The agency hoped Skylab would win back the support of at least a portion of the scientific establishment.

Initial Problems

The launch of the first *Skylab mission*, SL-1, took place on 14 May 1973, just a few months after NASA's final flight to the moon. There were problems from the start. A Saturn V rocket boosted the big cylindrical laboratory into Earth orbit. The lab, fashioned from the third stage of the Saturn rocket that carried it, like a man cave created in a shipping container, was unoccupied at this point. This point bears repeating. The Saturn V didn't just transport *Skylab*. A portion of the rocket actually *was Skylab*.

The four fan-blade-like solar arrays sticking out of the ATM gave *Skylab* its iconic windmill appearance. But the station was also designed to have two more large rectangular solar arrays, one on each side, as if it were gliding on a pair of stubby wings. Unfortunately, the station lost its micrometeoroid/solar-protection shield and one of its two big solar array "wings" as a result of mishaps shortly after takeoff. The second solar array appeared to be intact, but it failed to deploy—that is, spread out—as designed. These were crippling blows. *Skylab* was designed to operate largely on the electricity generated by its solar arrays. The micrometeoroid shield was intended to provide protection against random space debris crashing into the station, yes, but it was also designed to shield *Skylab* from solar radiation, keeping the station cool enough for human beings to live and work inside. As a result of these problems, temperatures aboard *Skylab* climbed to 130 degrees Fahrenheit. The station was uninhabitable under such conditions. *Skylab* seemed likely to be a total loss—an embarrassing and very expensive mistake.

SL-2, the first *crewed Skylab* mission, was originally scheduled to take off the day after Skylab 1 to ferry three astronauts to the station. Once NASA became aware of the damage to its new space habitation, however, the second launch was delayed for ten days as engineers and technicians scrambled to come up with ideas for fixing the crippled facility. Astronauts took to Marshall Space Center's giant swimming pool—the Neutral Buoyancy Simulator—to create the procedures that would be needed to remedy the problems. Another astro worked with the A. B. Chance Company of Centralia, Missouri, to procure a pair of utility worker's tools—a cable cutter and a "universal tool with prongs for prying and pulling"—both of which could be mounted on a three-meter pole. These tools and the pole were added to the mission's equipment manifest and carried into space when Skylab 2 launched on 25 May 1973. On 7 June, astronauts Pete Conrad and Joe Kerwin used the hardware while standing on the satellite as it cruised through the heavens, 270 miles above Earth, at seventeen thousand–plus miles per hour. The astronauts cut a piece of aluminum that was keeping the one undamaged solar array from deploying. Even after the aluminum strap was cut away, though, the recalcitrant contraption remained stuck, with one of its hinges essentially frozen in place. In a tense and wholly improvised procedure, Conrad and Kerwin had to use a rope to yank the array loose, after which effort the astronauts briefly soared off into space before being restrained by their tethers.

Skylab 2's crew also deployed a sort of Mylar parasol to shield the orbital workshop from solar radiation, a temporary fix that was improved a few weeks later when Skylab 3 astronauts Jack Lousma and Owen Garriott installed a sturdier twin-pole solar shield instead. Even with these repairs somewhat miraculously effected and the station restored to full functionality, the spacecraft was lopsided, like a mobile home with a broken window and an inelegant tarp that looked like a giant bandage. Remember, *Skylab* was essentially a repurposed Saturn booster fuel tank, with humans taking the place of several thousand gallons of liquid hydrogen. There was something star-crossed about the whole affair.

Trouble at Home

Space enthusiasts cite *Apollo 13* as the most spectacular example of a space repair. Justifiably so. There were three lives at stake, after all, and the world was watching. But the 1973 resurrection of the ailing *Skylab* by the crew of SL-2 surely comes a close second. Conrad's and Kerwin's heroics notwithstanding, though, the American public was largely unimpressed by heroics on the space station. It wasn't just that the country was still recovering from the techno-rave that was *Apollo 11*. *Skylab* simply wasn't pretty. The space walks of Gemini and the lunar salutes of Apollo were a thing of the past. The dazzling high-definition images captured by astronauts working outside the space shuttle were almost a decade in the future. To make it worse, the aims of this next phase of space exploration—solar observation and extended stays in low-Earth orbit—seemed fussy and diffuse by comparison with expeditions to the moon. The *Skylab* astronauts weren't *going* anywhere. Finally, there was other news occupying the nation's attention in the early seventies, and very little of it was good. There was the single biggest political scandal in U.S. history, for example.

The story of Watergate is way too big to be recounted in a history of the space program. Suffice it to say that it involved partisan skullduggery, burglary and break-ins, secret payments and strenuous cover-ups. Tape recordings made by the president of the United States in his own office—he had, in effect, bugged *himself*—revealed a venomous and bigoted behind-the-scenes Richard M. Nixon that few Americans cared to support. News also broke in 1973 that Nixon's vice president, Spiro Agnew, was facing a federal bribery investigation. Though he strenuously denied the allegations against him, Agnew eventually resigned from office, pleading guilty to a charge of tax evasion rather than face impeachment. The upshot of this series of unedifying revelations was a picture of the two leaders of the nation, Nixon and Agnew, as deceitful and shabby bad actors. "I am not a crook," Nixon declared, but few believed him. Watergate was a sinkhole in the country's image of itself as better than the godless communists, the bickering Europeans, the banana republics of Central and South America.

And the hits kept coming. The dollar was devalued. The comet Kohoutek, touted by NASA and the American press as a once-in-a-century cosmological spectacular, passed by the sun in late 1973 and early 1974 with barely a trace of a tail, disappointing millions of would-be comet watchers worldwide. While it certainly wasn't NASA's fault that the "hairy star"—for so the ancients called comets—failed to provide a show, the agency's breathless promotion of the event called its forecasting abilities into question. Native American activists occupied Wounded Knee, seeking redress for centuries of injustice, and American troops continued to return home from Southeast Asia, signaling the last stages of our bloody and disillusioning involvement in Vietnam. They were strange times in America, days of recrimination, self-doubt, and guilt. *Skylab*, it turned out, wasn't exactly the prescription for relief. And yet it was what we had. It was something rather than nothing. Astronaut Phil Chapman, selected to join NASA in 1967 as a scientist-astronaut, urged his colleagues to take seriously the opportunity to do field work in the sky—and to come up with good hard science for the astronauts on board the space station to do. "It is surely clear by now," he wrote, as planning for the orbital station began in earnest, that

> the titanic manned space program is not as unsinkable as it may have seemed a year ago. We have very few friends left, in the scientific community, in Congress, or amongst the general public. . . . It is a difficult time, when science and technology are depicted as destroyers of the environment and/or instruments of war and the nation's attention is engrossed in urgent problems such as Vietnam and pollution and the

ghettos. All that means is that we must perform better and fight harder instead of meekly going under, crying excuses.

Skylab Science

Aside from being a way to keep Apollo technology and personnel online, *Skylab* was also an obvious response to the Soviets' crewed space station, *Salyut 1*, which was deployed in 1971. Three cosmonauts visited the station in June of that year and stayed for twenty-three days, setting a record for the length of their stay in space. The crew didn't have long to enjoy their achievement, though. During their return to Earth, their Soyuz spacecraft malfunctioned. The resulting depressurization of the vessel killed all three men. It is still the worst tragedy to strike a Soviet or Russian space mission.

In the competitive sense, *Skylab* was a quick success. "SKYLAB CREW TAKES OVER SPACE ENDURANCE RECORD" crowed one American daily on 19 June 1973, after the first *Skylab* crew's stay in space surpassed the Soviet mark. But that was just the start. Altogether, three crews of three astronauts visited the American space station in 1973 and 1974, spending a then-staggering total of 171 days in orbit. The first crew, that of Skylab 2, inhabited the station for 28 days. The crews of Skylab 3 and 4 lived there for 59 and 84 days, respectively. These totals seem paltry now, but only because *Skylab* astronauts paved the way for their successors' achievements. More recently, Scott Kelly spent 340 days in orbit on our current orbital hang-out, the International Space Station, or ISS. Christina Koch set a women's record, logging 328 days on the ISS in 2020. And cosmonaut Valery Polyakov spent a whopping *437* days in Russia's *Mir* space station in 1994 and '95. Polyakov's good health and stable mental outlook when he returned to Earth have been cited as evidence that extended journeys in space are feasible. Back in Russia, he walked away from his reentry-scorched Soyuz capsule and proclaimed, "We can travel to Mars."

The nine astronauts who visited *Skylab* on missions 2, 3, and 4—and that first crew in particular—spent a significant amount of time repairing it. The rest of their time was meant to be devoted to science. The nation's high school students competed to get their proposals on board, and the best of these were incorporated into *Skylab*'s task list. The crew monitored and recorded solar activity, capturing important information about solar flares and other phenomena produced by our local star; indeed, data gathered by *Skylab* contributed to the work that won scientist Richard Giaconni a Nobel Prize in Physics in 2002 for studies of X-ray astronomy.

The astronauts were themselves the subject of one of the station's central experiments, a fine-grained survey of changes in human physiology in space. NASA tried to make the spacecraft comfortable for the extended stays the crews had signed up for. There was an exercise bike to keep hearts and muscles strong, a not-very-efficient hot-water shower, and a wardroom for communal dining, along with three chambers for sleeping, reading, and general relaxation. The orbital workshop proved to be a huge asset. It looked a lot like a massive enclosed culvert in space, and functioned as a combination laboratory, test bed, and jungle gym. The astronauts used it to demonstrate the strange behavior of liquids, solids, and grown men in micro-gravity. They ran on walls, somersaulted in flight, lifted each other with a single finger, and so on. Some of the amenities on the vessel worked fine. Others didn't. The shower was a flop, for example, but the viewing portal in the wardroom turned out to be a big hit. It was all part of the experiment. Lessons NASA learned during the *Skylab* missions, like the importance of physical exercise during space travel, are now reflected in everyday life on the ISS.

So what *do* we know about the effects of space on the human body? Now that human beings have occupied low-earth orbit for sixty years, we have a list. First, some astronauts get sick. Space adaptation syndrome (SAS) is a mysterious but common malady that affects many spacefarers, especially early on in their missions. Symptoms include disorientation, headaches, and nausea—in some cases, *severe* nausea. Astronaut Rusty Schweickart was hit with such a bad case of SAS on *Apollo 9* that it may have disqualified him from a lunar landing mission.

A less immediate but more serious health issue relates to the fact that because astronauts are outside the protective atmosphere of Earth, they're exposed to higher doses of radiation. While relatively short-term exposure seems not to have dramatically increased the risk of cancer among space travelers, the jury is still out on how big a danger this radiation risk is and for how long it can be tolerated. Another problem: muscles grow weaker and bones lose density in space because gravity is not making the body work to counteract its effects. As a result, astronauts on the ISS are required to spend two hours a day exercising to stay strong.

Another common consequence of space travel is that, particularly early in a flight, an astronaut's head and upper body generally will become bloated, or "puffy." Astronauts broadcasting from space tend to look like they've just gone a few rounds with a hard-punching middleweight from a Pennsylvania coal town. This is because gravity is no longer pulling blood and other bodily fluids down, so such fluids accumulate in the upper body. The mucous membranes of the nose swell, so astronauts often have congested noses. The fluids in the body eventually balance out, and facial swelling typically begins to disappear after a few weeks. But one effect of this upward fluid migration is a little more troubling. Pressure from fluids in the head can distort the shape of the eyeballs, causing what's known as spaceflight-associated neuro-ocular syndrome, or SANS. This in turn can lead to long-term vision problems for the astronauts—definitely a liability in an environment where spacefarers are constantly checking monitors, checklists, and equipment as part of their jobs. And while the risk of degraded vision may be acceptable for short-term missions, no one can say for sure what will happen on a mission that lasts, say, three years, which is about the time it will take for a crew to get to Mars and back using currently available technology. Needless to say, research continues.

A Gift for Environmentalists

Along with observation of the sun and physiological monitoring of the astronauts, *Skylab* had a third major focus: Earth. Using a variety of cameras, the astronauts captured a stunning number of images of our globe. As a NASA publication puts it, "The more than 40,000 photographs made of the Earth and the thousands of observations recorded on miles of magnetic tape provided a mass of data . . . of great value to those involved in improvements of agriculture and forestry, geological applications, studies of the oceans, coastal zones, shoals and bays, and continental water resources, investigations of atmospheric phenomena, regional planning and development, [and] mapping and further development of remote sensing techniques." Taken under the rubric of the Earth Observation Program, these images have ever since provided an important baseline for examining environmental and geophysical changes on our home planet. Due perhaps to the nation's post-Vietnam fatigue, the astronauts were told not to call the subjects of their photographs "targets," as the word sounded too militaristic. And because it was a period of détente, an easing of hostilities in the long war between the West and the East, they were also prohibited from taking pictures over the USSR and China, though they weren't, of course, restricted from looking. SL-3 pilot Jack Lousma remembers gazing with particular interest down toward the Soviets' highly restricted Baikonur launch complex.

In gathering these images, the *Skylab* astronauts laid the groundwork, so to speak, for one of the most important of NASA's scientific endeavors—the continuing study, not of space, but of Earth. Many writers credit *Apollo 8* with jump-starting the American environmental movement by capturing images of our globe in a single frame, dwarfed by the immensity of the cosmos. The clear message was that our home planet is a lonely miracle, deserving of attention and care. The *Skylab* missions were in a sense an acknowledgment of that new perception. The space station came online as the environmental movement was reaching full strength. Ecological angst filtered into popular culture. In the 1972 film *Silent Running*, a scientist sacrifices himself to save a fragile, space-grown biosphere from the callousness of corporate bureaucracy. The sci-fi actioner *Soylent Green* hit theaters a year later, right around the same time *Skylab* was launched. Wielding a wafer-thin plot about a giant corporation turning an unknown substance into food and dispensing it to a starving populace, Hollywood forcefully warned Americans about the hazards of pollution, climate change, and deceptive labeling. (Spoiler alert: "Soylent Green is PEOPLE!") In the movie, the character Sol Roth, played by the great Edward G. Robinson, explains how things got so bad:

> When I was a kid, food was food. But our scientific magicians poisoned the water. Polluted the soil. Decimated plant and animal life. Why, in my day, you could buy meat anywhere! How can anything survive in a climate like this? A heat wave all year long. The Greenhouse Effect. Everything is burning up.

Influential books like Paul Ehrlich's *The Population Bomb*, E. F. Schumacher's *Small Is Beautiful*, and Edward Abbey's *Desert Solitaire* struck notes of admiration for the natural world and anxiety about the ways in which humanity and technology were screwing it up. The chorus of concern eventually resulted in change. In December of 1973, as the third crewed *Skylab* mission orbited overhead, Congress passed the Endangered Species Act, arguably the most important piece of environmental legislation in the nation's history. It therefore seems appropriate that the three *Skylab* crews spent considerable time photographing, and indeed just *gazing at*, the blue jewel passing beneath them.

The agency's interest in earth science has never really subsided, though sometimes it seems as if the public's has. While the scientific freak flag that was *Skylab* is long gone, NASA continues to provide some of the best and most complete atmospheric and oceanic data about our planet. And the agency doesn't mince words. "HUMAN ACTIVITY IS THE CAUSE OF INCREASED GREENHOUSE GAS CONCENTRATIONS" says the headline on one of its website pages. Other text advises that since systematic scientific assessments began in the 1970s, "the influence of human activity on the warming of the climate system has evolved from theory to established fact." Such assessments can be seen as a direct legacy of *Skylab*.

Less obvious but at least arguable is the proposition that *Skylab* restored America's admiration for science after years in which its work, as embodied by "advances" like DDT, Agent Orange, napalm, and nuclear weaponry, was seen primarily as a tool chest for military and industrial interests, indifferent to the health of the human race and its home planet. For that, if nothing else, we owe the project and its astronauts a debt of gratitude.

No, *You're* a *Skylab*

Skylab is perhaps best known today for its unfortunate demise. It fell out of the sky. Truth be told, lots of things fall out of the sky. Most of them, though, are fairly small and burn up as they enter earth's atmosphere. *Skylab* was the size of an army barracks. The "decay" of the station—that is, its gradual loss of altitude—as the decade wore on was inevitable, and indeed well documented. NASA wasn't worried at first, because the agency anticipated that something called the "space shuttle" would be online in time to travel to the space station and boost it back up into a longer-lasting trajectory. As late as 1978 there was talk of having the shuttle visit and reactivate the station for renewed habitation. It never happened. Delays in production of the shuttle dragged on, and unusually high solar activity heated the earth's atmosphere just enough to slow *Skylab*'s orbit, which in turn increased its rate of orbital decay. The breakup of a nuclear-powered Soviet satellite over northern Canada in January 1978 gave people around the world a preview of what was coming: fiery wreckage raining from the skies.

The resulting low-level panic about the space station's unpredictable reentry path saw entrepreneurs selling bottles of something called "*Skylab* repellent." Several residents of England hid in a cave, while the *San Francisco Examiner* offered a $10,000 reward to the first person who could find and deliver to the newspaper's office a bona fide piece of the fallen satellite. America's reputation for engineering excellence hit a new low when a man was reportedly killed in Indonesia after he offered another man the ultimate insult, calling him a "Skylab." The hysteria was partly humorous, to be sure, but it wasn't entirely misguided. The artificial asteroid about to hit Earth was a whopper. In fact, *Skylab* was the largest single pressurized structure ever to be placed in orbit. The errant space station entered Earth's atmosphere on 11 July 1979, breaking up a mere ten miles above the planet's surface—much lower than predicted. No injuries were attributed to the fall of the seventy-seven-ton station, but property and physical damage were certainly possible when the descent occurred and *Skylab* sprinkled itself like flakes of an unwanted condiment over portions of western Australia.

So, *Skylab*: our Appalachia in the sky. By and large, the nation's first space station failed to produce America's most memorable moments in the cosmos. But it was still spaceflight—daring and risky and always demanding. Time has allowed us a heightened appreciation of *Skylab*'s achievements, quirks and all. Perhaps the program's most important accomplishment was its creation of a whole new field of study for NASA. This occurred when the men in the pumpkin-colored flight suits turned their telescopic lenses in the other direction—back, that is, toward the improbable solar-powered oasis called Earth. So what if no one was paying attention? In late January of 1974 television networks broadcast in prime time the weigh-in of heavyweight boxers Muhammad Ali and Joe Frazier for their Madison Square Garden title fight, which Ali won on points on 28 January. By contrast, the 8 February splashdown of the crew of Skylab 4, the final *Skylab* mission, was largely ignored. It received no television coverage at all. Julie Gibson, wife of solar physicist and *Skylab* astronaut Ed Gibson, was as puzzled as she was disappointed.

"Don't you think that what Ed did is as important as two prize fighters weighing in?" she asked journalist Molly Ivins in the pages of the *New York Times Magazine*. "Don't you?"

For the record, Julie: *Yes. We do.*

In Which All Is Temporarily Forgiven

One more flight needs a shout-out. The Apollo-Soyuz Test Project (or ASTP) was a one-off American-Soviet mission in 1975 meant to mark the end of the space race and usher in an era of friendlier relations between the United States and the USSR. America's astronauts left Earth using Apollo hardware. In fact, the ASTP is sometimes semi-seriously referred to on our side as "*Apollo 18.*" The Soviets flew their Soyuz craft, and the two vehicles, *Soyuz* and the Apollo command module, met up in orbit using a specially designed collar that allowed the two crews to go back and forth between the spaceships. The three spots on the American mission were allotted to Tom Stafford, Vance Brand, and the grizzled Deke Slayton, finally cleared for spaceflight after a long medical suspension for heart issues. The mission went well. Linked together, the space vessels looked like two bugs kissing through a harmonica. Our astronauts shook hands with their cosmonauts, Alexei Leonov and Valeri Kubasov. The combined crew made toasts and constructed a commemorative plaque. It was sorely needed proof that the old enemies could actually work through their differences and cooperate in the difficult business of space exploration—or at least space fraternization. It was also the high-water mark for Soviet-American relations.

Unfortunately, the tide soon ebbed.

NASA's Eleven Coolest Astronauts

1. Neil Armstrong: Humanity's first moon man was immaculate and indestructible. No one should rest until there is an opera written about him. In heaven.
2. John Glenn: That's *Senator* Glenn to you, by the way. Everybody's All-American—as straight as a skyscraper, polite as a preacher, a killer in the cockpit. Our first Earth orbiter, he took a short break to go to Washington and came back thirty-six years later to fly on the shuttle.
3. Sally Ride: NASA's first woman in space. A civilian in what had previously been a military man's world, she handled industrial-strength pressure and a shuttle-load of ridiculous questions with composure and the occasional sharp-edged retort. She was a glacier, until she wasn't. Then she was a scrapper who never walked away from a fight.
4. Mike Collins: NASA's Renaissance man, gentle, bright, and slyly funny. Author in 1974 of *Carrying the Fire: An Astronaut's Journeys*, winner and still champion of the Astronauts' Books, Heavyweight Division.
5. John Young: Put a movie-star face on a quirky, shade-tree mechanic and then watch him set time-to-climb records in a modified Phantom fighter jet at the age of thirty-two. As for cool? The Georgia drawl and down-home witticisms help; so do six successful space missions, including a moonwalk on *Apollo 16* and command of the first flight of the space shuttle. And then, of course, there's the pipe and the turtleneck sweater . . .
6. Ron McNair: Musician, physicist, karate black belt, and optimist, McNair sometimes seemed like he was playing himself in a musical

called *McNair!* No one else has ever appeared quite as joyful to be flung into the cosmos.

7. Al Shepard: The first American astronaut in space later commanded an Apollo lunar mission as well. Had the world on a platter and sent it back for more sauce. Supreme confidence. Or was it arrogance?

8. Eileen Collins: Gender pioneer, loving mom, and stone-cold pilot, former air force officer Collins was the first woman to command a space shuttle mission. What's in that clutch, you ask? Authority. But she only takes it out when she needs it.

9. Kathy Sullivan: The nation's first female spacewalker—and, years later, the first woman to visit Earth's deepest cleft, the Pacific Ocean's Mariana Trench. A first-class mind, adventurous, capable, and blunt.

10. Jim Lovell: The four-time spacefarer survived fourteen days in a capsule the size of a Volkswagen with astronaut Frank Borman, but he's best known for his role in bringing *Apollo 13* back home after a craft-crippling explosion. The 1995 movie version of the mission, starring Tom Hanks, is widely credited with rekindling public affection for NASA. Houston's Han Solo, now ninety-six, remains lucky, loquacious, and unflappable.

11. Hoot Gibson: One of NASA's best-ever pilots, equipped with just the right combination of intelligence, technical savvy, and complete insanity. When Gibson went to fly, the weather called for a report on *him*. Air ace born a few years too late for Apollo; he might have rivaled Armstrong.

1. Brilliant and focused, Robert Goddard was a rocket scientist before such things existed, an American original equally at home at a lathe or in a library. Here he's seen just before launching his first liquid-fueled rocket in March 1926. Courtesy of NASA.

2. Dr. James Pickering, Dr. James Van Allen, and Wernher von Braun hold up a replica of the Jupiter-C rocket that launched America's first satellite, Explorer 1, into space in January 1958. Courtesy of NASA.

3. One thing the early astronauts shared was confidence. Here, Mercury astronaut Scott Carpenter is suited up for his five-hour flight on Mercury-Atlas 7 in May 1962. Courtesy of NASA.

4. (Opposite top) Seen here at work at Cape Canaveral in the early sixties, Chris Kraft was a flinty Virginian who, more than anyone else, invented Houston's mission control facility and later presided over it with absolute authority. Courtesy of NASA.

5. (Opposite bottom) Frank Borman pauses to take in some sunshine after completion of his fourteen-day *Gemini 7* space voyage, which he accomplished in December 1965 with fellow astronaut Jim Lovell. Conditions in the cramped capsule were less than ideal. Courtesy of NASA.

6. Astronaut Dave Scott peers out of his command module during the *Apollo 9* mission in early 1969. Courtesy of NASA.

7. Buzz Aldrin salutes the American flag on the lunar surface on 20 July 1969, capping the greatest voyage in history. Courtesy of NASA.

8. Al Shepard practices for his 1971 *Apollo 14* moonwalk with fellow astronaut Ed Mitchell. Courtesy of NASA.

9. In a photo taken not long after launch in 1973, NASA officials discuss problems associated with *Skylab*'s missing micrometeoroid shield. From left to right, the group includes Jack Kinzler, whose *Skylab* sunshield solution earned him the NASA Distinguished Service Medal, along with William Schneider, ace spacecraft designer Max Faget, Dale Myers, JSC director Chris Kraft, and Kenneth Kleinknecht. Courtesy of NASA.

10. In July 1975 American astronauts rendezvoused with Soviet cosmonauts in the Apollo-Soyuz Test Project. Here, the famously gruff Deke Slayton poses for a portrait with Soviet cosmonaut Alexei Leonov. Courtesy of NASA.

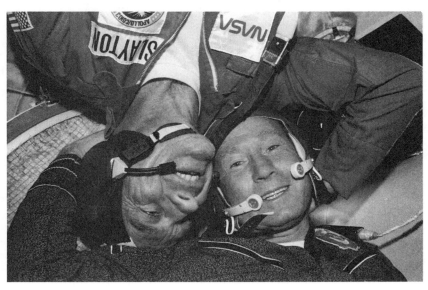

11. Astronaut Anna Fisher joined NASA as a member of the 1978 class, the "Thirty-Five New Guys," NASA's first big attempt to bring diversity to the astronaut corps. Fisher, a physician, became the fourth American woman to reach space when she flew on STS-51A in 1984. Courtesy of NASA.

12. In February 1984 astronaut Ron McNair filmed portions of STS-41B for a documentary titled *The Space Shuttle: An American Adventure*. Crewmembers called him Cecil B. McNair in a reference to legendary filmmaker Cecil B. DeMille. Courtesy of NASA.

13. Astronauts Carl Meade and Mark Lee (with red stripes on his pressure suit) test the little "rescue jetpack" called the "self-assist for EVA rescue" device (SAFER) on STS-64 in September 1994. Courtesy of NASA.

14. Unassuming and unflappable, astronaut Eileen Collins was both the first woman to pilot a space shuttle mission (on STS-63 in 1995) and, in 1999, the first female to *command* a shuttle mission (on STS-93). Courtesy of NASA.

15. In 2020 the SpaceX Crew Dragon capsule ended an embarrassing and expensive American reliance on the Russian space program to get astronauts to and from the International Space Station. Courtesy of NASA.

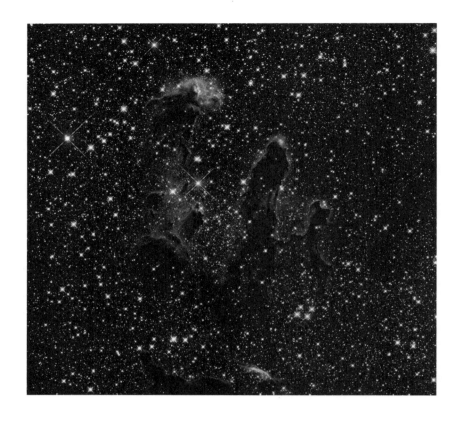

16. This 2014 view of the so-called Pillars of Creation, clouds of gas and dust conducive to the formation of stars, is one of the Hubble Space Telescope's best-known images. The finger-like pillars are some seven thousand light years away from Earth, in the Eagle Nebula. Courtesy of NASA.

17. The crew of *Artemis 2*, currently scheduled for launch in late 2025, from left to right: pilot Victor Glover, commander Reid Wiseman, mission specialist Jeremy Hansen (a Canadian), and mission specialist Christina Koch. *Artemis 2* will perform the first crewed lunar orbit since the Apollo era. Courtesy of NASA.

18. NASA's *Cassini* probe imaged Saturn's moon Enceladus, which is thought to have vast oceans of water beneath its thick crust of ice. Some scientists believe these oceans may harbor life. Courtesy of NASA.

12

Probes, Rovers, and the Golden Record

Several American space probes launched in the seventies
have proven to be remarkably productive—and long-lived.

Although no American entered Earth orbit between 1975 and 1981, the seventies were still busy years for space exploration. Indeed, the decade saw the launch of a number of ambitious probes and satellites, both at home and abroad. The Soviets landed two rovers on the moon and put probes on Venus, though the probes didn't last long in the planet's hellish atmosphere. NASA's *Mariner 9* became the first spacecraft to orbit Mars in 1971 and sent back photographs that revealed a far more varied and interesting planetary surface than astronomers had previously suspected. Among *Mariner 9*'s discoveries was the Valles Marineris, a massive declivity that runs like a scar 2,500 miles across the planet and descends as deep as 4 miles below the Martian surface. Twenty-five hundred miles is about the distance between Las Vegas and Philadelphia. As dramatic as it is, Arizona's biggest tourist attraction, the Grand Canyon, only counts as the "Pretty Good Canyon" next to the epic Valles Marineris.

Pioneer 10 and *Pioneer 11* left Earth in 1973 and 1974, respectively. Both of the probes visited Jupiter, while *Pioneer 11* went on from there to become the first spacecraft to reconnoiter Saturn. The machines are still out there—*way out there*—and flying, though they have long since ceased transmitting communications back to Earth, so we're no longer sure of their exact locations. The *Pioneers* each carry a gold-anodized aluminum plaque engraved with pictorial greetings to far-flung civilizations. The astronomer Carl Sagan and his second wife, Linda Salzman Sagan, were prime movers in the effort to include these high-tech hellos. Among the illustrations they chose to include are a chart of Earth's place in the solar system and, within it, the trajectories of the *Pioneer* spacecraft. Also included are line drawings of two human beings, one male and one female, naked and apparently contemplating their new alien acquaintances with equanimity, if not much enthusiasm. The figures attracted considerable attention at the time and—perhaps inevitably—a fair amount of cross-cultural grumbling. The figures were faulted, for example, for being too blond and too white. They were also panned for being too African (the noses are supposedly flatter and wider than the average Caucasian proboscis) and too Asian (some people claim that the woman's eyes have an epicanthal fold). Some say they're too naked. Others have opined that they're not naked enough: the male, for example, appears to be wearing a small horseshoe over his groin, while the female has been neutered, her nether parts Barbiecized and thus rendered the subject of speculation. Both are clean shaven—certainly an option, but not necessarily representative of humanity as a whole. But perhaps these are quibbles. The Sagans at one point thought about having the couple hold hands. They eventually realized that this might be mistaken by some distant alien as a representation not of two human beings but rather of a single, much larger, six-limbed organism. Sending an intelligible message across the universe is harder than it might seem.

In 1972 the United States launched the first of its long-running series of Landsat satellites (first called the Earth Resources Technology Satellite, or ERTS) from California's Vandenberg Air Force Base. Landsat was the first satellite specifically designed to capture high-definition photographs of Earth's oceans and land masses. Its developers and proponents were interested in applying space science to "pressing natural resources problems being compounded by population and industrial growth," as secretary of the interior Stewart Udall said in 1966. The program is currently managed by the U.S. Geological Survey, which recounts its origins this way:

> The DOI, NASA, and the U.S. Department of Agriculture (USDA) ... embarked on an ambitious effort to develop and launch the first civilian Earth observation satellite. These revolutionary satellites would be set in a heliosynchronous, near-polar orbit, completing several revolutions around the Earth every day to capture the land surface of the planet. The heliosynchronous, near-polar orbit means the satellite passes near the North and South pole consistently as it revolves around the Earth [and] allows a Landsat spacecraft to pass over the equator at a different longitude on each revolution, resulting in the spacecraft completing a full image of our planet after 251 orbits, about 16 days.

Currently, the eighth Landsat (Landsat 9, since one of the series, Landsat 6, failed to deploy) is in operation. The satellites have amassed some ten million images over the course of fifty years, capturing changes over time to ecosystems around the world, from rainforests to African deserts, from Lake Baikal to the Chesapeake Bay. Since 2008, Landsat imagery has been available for download at no charge from a variety of data portals. In fact, you've probably used some of this data yourself, as Google has employed Landsat images for years in its Google Earth and Google Map applications.

Vikings and *Voyagers*

Despite the success of *Mariner 9*, the Mariner program featured just one more mission, the *Mariner 10* visit to Mercury in 1973. Nevertheless, NASA was able to use technology developed for Mariner in constructing and launching what came to be called the *Viking* probes in 1975. In those days it was the agency's practice when launching uncrewed spacecraft to send two of the same type. The thinking was that if one should fail, the other would carry on in its place. Each of the two *Viking* probes consisted of two components, an orbiter and a lander. Each lander separated from its associated orbiter after a number of trips around the planet and then descended to the surface of Mars.

Viking 1's lander touched down on the Red Planet on 20 July 1976—seven years to the day after *Apollo 11* reached the moon. *Viking 2* followed on 3 September. The *Viking* landers transmitted the first detailed images taken on Mars, monitored weather conditions, and listened for Mars-quakes and other tectonic events. But perhaps the primary goal for each lander was to sample Martian soil for evidence of life. Many astronomers had grown convinced in the early years of the twentieth century that the planet was scored by canals, and that this in turn indicated the presence of intelligent beings. Aggressive beings? With eye-melting laser pistols? Even in the seventies, such sci-fi fever dreams had long since been relocated to well outside the solar system. Even so, scientists and laypeople alike retained a lingering hope that Mars nurtured at least some rudimentary life, either in the present or at some point in the past.

While the data collected by the *Viking* probes were largely viewed at the time as indicating that no life existed on Mars, the results are still being studied. Indeed, an increasing number of scientists are considering the possibility that researcher Gil Levin's "labeled release" experiment, hosted on the probes, did in fact find evidence of biology in the Martian soil. Whatever the ultimate outcome of that debate, other scientists are interested in whether life could have existed long ago, or whether it did or might still exist below the surface of Mars. The mission was at any rate a huge technical success, with launch vehicles, orbiters, and landers performing at or well above expectations. And while the landers may have failed to find life at the two locations where they touched down, they did return compelling evidence that Mars once held vast amounts of water. Indeed, some of this water remains, chiefly in the form of ice at the planet's poles.

As spectacular as *Viking*'s visits were, they were just a prelude to an even more remarkable mission. NASA launched two *Voyager* probes from Cape Canaveral in 1977, taking advantage of a rare alignment of the solar system's four largest planets. This convenient orbital parade takes place every 176 years and presents a mouth-watering prospect for solar system explorers—almost as if terrestrial tourists were to find that New York City, Rome, Istanbul, and Beijing were all suddenly accessible on a single high-speed rail line. The *Voyager* probes used a procedure called "gravity assist," in which they employed the gravitational pull of the big planets not only to draw them closer, but also then to sling them on their way toward their next destinations. Both of the *Voyagers* visited Jupiter and Saturn. *Voyager 1* also scanned Jupiter's moon Io, spotting volcanic activity on the moon's surface. *Voyager 2* soared off after its flyby of Saturn to take photographs and measurements of Uranus and Neptune, thus becoming the first and still the only probe to visit those planets. Together, the *Voyagers* discovered a total of ten new moons and captured stunning evidence of volcanic activity and the presence of extensive ice crusts on orbs long thought to be as arid and empty as our own ghostly Luna. The effect of these discoveries can hardly be overstated. It was as if one were to walk into a museum of natural history, flick on the overheads, gaze at the row of preserved heads on the wall, and then see the eyes on one of the heads blink in the light. Our sister planets are *alive*, geologically speaking—alive and changing and still writing their own stories.

A discussion of volcanoes on a Jovian moon 444 million miles away from us is probably as good a time as any to ask a nagging question related to space exploration. Who cares? It's easy to understand why we're curious about the moon and Mars. We might be able to live there someday. In fact, we might *have to* live there someday. And sure, let's study asteroids. They're loaded with minerals—and some minerals are extremely useful, which is another way of saying *valuable*. In addition, asteroids have a nasty habit of periodically slamming into our planet, causing annoying phenomena like mass extinctions. So by all means—map the asteroids! And of course we need to know about the sun, because solar behavior directly impacts our activities here on Earth and in Earth orbit. But Jupiter? Alpha Centauri? Distant black holes? Why spend time and money trying to figure out how distant planetoids were formed?

Of course the same crabby questions could have been asked of Galileo when that noted Florentine troublemaker first trained his homemade telescope on Jupiter's moons four hundred years ago. Why bother? Especially when bothering is about to get you crossways with the powers that be, who have already decided they know what matters and what doesn't and have shown they are willing to yank out the entrails of those who disagree. But some of us—not the authors, necessarily; we are Netflix junkies—can't help it. Why, Galileo wondered aloud, were there moons revolving around Jupiter? Conventional wisdom held that Earth was the center of the universe and that all the spheres one could see in the heavens revolved around *us*, for whom Creation had been expressly fabricated. Such explanations sufficed at the time. They seemed to make sense. But one thing science shows us is that we can't always know what's useful and what's not. A discovery that seems arcane and pointless—indeed, offensive—in 1610 might become very significant a few years later. We now think that Galileo's work was extremely useful. While the Catholic Church kept the sun-centric theories and observations of Copernicus, Galileo, and others on its forbidden list well into the nineteenth century, considering such writings as a sort of cosmological pornography, their work started us on the road to understanding how our solar system really works. They led, slowly but surely, to Neil Armstrong's walk on the moon and Elon Musk's dreams of establishing habitations on Mars, to the *Voyager* probes and our current studies of dark energy and exoplanets, gravitational waves and the formation of stars. It's impossible to say where these interests will take us in the future and what we'll learn along the way.

Aside from personal or species or planetary self-interest, we human beings want to figure out how things work and why they are the way they are because curiosity is hardwired into our genome. We want to know because whatever we study, we are studying *us*. We want to know because energy and matter, here and a hundred light years away, are different faces of the same reality. And the odd and far-flung facts we collect, as seemingly random and isolated as they are, are clues to the same ultimate riddles. Are we all just bugs caught in the folds of a giant cosmic accordion, forever expanding and contracting in sequence? Or are we castaways on a lonely island surrounded by a universe that is rapidly pulling itself apart? What is the nature of this vast dark sea we're swimming in? What, if anything, is our purpose? Will we ever know? *Can* we ever know?

Not if we don't ask. And so we launch our mechanical question marks out into the darkness.

The Golden Record

Notably, each of the spindly *Voyager* probes carries a phonograph record that contains images and sounds meant to introduce Earth and its residents to any aliens who might happen upon the probes. For those advanced extraterrestrial civilizations that might have moved on from LPs to eight-track tapes, graphic instructions propose the correct way to access the information contained on the disk. And unlike Rolling Stones albums, each record cover contains a small amount of uranium-238 with a radioactivity of about .00026 microcuries. The rate of decay of this material could be used by aliens to determine the amount of time the disk—and thus the satellite that carries it—has been traveling. The disk, often called the Golden Record, is made of copper coated in gold and protected by an aluminum cover. It carries photos of our planet and its life forms, a range of scientific information, and a medley, "Sounds of Earth," that includes the sounds of a baby crying, whales calling to each other, and the slow pulse of waves breaking on a shore. Also included is a collection of music with works by Mozart, an Australian aboriginal chorus, Bulgarian folk singer Valya Balkanska, and rock-'n'-roll icon Chuck Berry, performing "Johnny B. Goode." Finally, the record contains greetings in fifty-five languages. Perhaps most notable is this poignant salutation recorded by Kurt Waldheim:

> As the secretary general of the United Nations, an organization of the 147 member states who represent almost all of the human inhabitants of the planet Earth, I send greetings on behalf of the people of our planet. We step out of our solar system into the universe seeking only peace and friendship, to teach if we are called upon, to be taught if we are fortunate. We know full well that our planet and all its inhabitants are but a small part of the immense universe that surrounds us, and it is with humility and hope that we take this step.

A year after the *Voyager* probes left Earth, comedian Steve Martin appeared on *Saturday Night Live*. Playing a cornball psychic, Martin predicted the response of the aliens who would one day receive the Golden Record. The four words proved, he said, that intelligent life exists elsewhere in the universe. Their message? SEND MORE CHUCK BERRY.

Into the Black

After their planetary flybys, the *Voyagers* just kept . . . *voyaging*. They outpaced their technological cousins, *Pioneer 10* and *11*, some years ago and are now on their way to distant stars. As physicist David Bohlmann points out, as of this writing, *Voyager 1* is more than fifteen billion miles from Earth and transmitting data back to us by means of a whispery signal that takes over twenty-two hours to get here—longer than the length of time Neil Armstrong and Buzz Aldrin were parked on the moon. Astonishingly, even as it closed in on fifty years of service, *Voyager 1* continued to provide valuable data. Most recently this has involved information regarding the heliosphere, the big bubble of solar wind within the solar system. *Voyager 1*'s findings have confounded traditional understandings of how the heliosphere is configured and how it works. As environmental historian Steven J. Pyne put it, speaking of the two probes as one mission, "Voyager did things no one predicted, found scenes no one expected, and promises to outlive its inventors. Like a great painting or an abiding institution, it has acquired an existence of its own, a destiny beyond the grasp of its handlers. The mission . . . continues; the spacecraft, so obviously a piece of engineering, endures as art, a project sold as science persists as saga."

13

The Butterfly and the Bullet

America's first reusable spacecraft, the space shuttle was sophisticated, versatile, and stylish. It promised the world. It flew for thirty years, and it broke a million hearts.

Unless you were involved with launching and monitoring interplanetary probes or at the track betting big on a three-year-old colt called Secretariat, the seventies were pretty awful. Popular music seemed to ooze rather than resonate. Men wore plaid-patterned bell-bottoms. A rolling bomb called the Ford Pinto was the country's most popular car, and Americans flocked to see cinematic gems like *Benji*, *Herbie Rides Again*, and *Freebie and the Bean* at their local theaters.

The early seventies also marked rock bottom for the space agency's budget. In announcing his "vision" for NASA's future in January of 1972, President Nixon killed a constellation of dreams. There would be no missions to Mars or additional visits to the moon, no massive space wheels twirling in Earth or any other orbit—not any time soon, at least. Nixon had delighted in the successes of Apollo and used the program as best he could for political and promotional means. Indeed, in one of the mildly bawdy parody films the astronauts produced in the late sixties, one wag proposed repurposing Neil Armstrong's famous quote regarding one small step for man to add the punchline, "One more term for Nixon." The California Republican was savvy enough to realize that voters were no longer interested in NASA's grand ambitions. Nevertheless, cognizant of how useful it might be to promise jobs and investment during an election year, he directed the agency to focus on finishing up plans for a "space plane" that could travel back and forth from Earth to a space station that would—*might?*—be constructed at some point in the future. In the federal budget that was finally approved in August of that year, Congress approved $200 million in funds for development of the vehicle. Thus was America's Space Transportation System given its official green light. That was its formal name, anyway—the Space Transportation System, or STS. The public just called it the space shuttle. It was a vehicle created by committee. It was a beautiful bird with concrete feet, complicated, compromised, ambitious and expensive. No spaceship has ever inspired as many arguments. At this point in history, though, the space shuttle's iconic winged orbiter is the vessel that soars through most people's minds when they think about cosmic exploration.

While early plans envisioned the shuttle being carried aloft by a piloted, winged booster vehicle, which could be reused any number of times, budget constraints led to the development of a more mundane alternative: launch from a pad, with an expendable fuel tank and reusable boosters. Even so, the project was expensive. To justify the costs, Washington mandated that the shuttle be considered the go-to transportation source for not only civilian missions but also those of the military. While the Pentagon reluctantly acquiesced, its requirements led to an increase in the orbiter's payload bay size and the redesign of its wings to bolster the craft's cross-range capability—its ability to glide greater distances once it returned to Earth's atmosphere. Such changes resulted in a heavier, more complex vehicle than NASA by itself would have created.

Despite such compromises, the shuttle held considerable promise. Its backers hailed it as a safe, economical, reusable space jet. It could take off from pads at California's Vandenberg Air Force Base or from Kennedy Space Center in Florida. It would reliably deploy and service satellites, military and civilian, and it would cut the costs of payload launch and delivery dramatically. It would, in short, be all things to all people.

As familiar as the outline of the shuttle orbiter looks today, in the seventies, the vehicle—with its large external fuel tank and two solid rocket boosters attached like cosmic torpedoes—seemed deliciously futuristic and sleek. And indeed, the shuttle was a step and a half forward in aerospace engineering. Veteran designer Max Faget had started spitballing ideas for the new vehicle in 1969, crafting a concept out of balsa wood and paper. He developed a working model two years later. "PLANNED SPACE SHUTTLE TRANS-PORT CALLED REVOLUTIONARY" read the headline of one 1971 wire-service story. The orbiter was the "space plane" aerospace engineers and sci-fi nerds had been expecting for years. While it couldn't take off from a runway and soar into low-Earth orbit, as some hoped—engineers are *still* working on that project—the orbiter was extraordinary nevertheless. It could seat eight astronauts. No previous spacecraft had left Earth with more than three. The orbiter could carry over fifty thousand pounds of payload. Previous piloted spacecraft were unable to haul much more than their own crew and equipment, plus a few souvenirs and pop-top tins of turkey and gravy. The shuttle could maneuver in space, survive the atmospheric volcano of reentry, and glide home to a landing in California, Florida—or, in one case, New Mexico.

NASA awarded the contract for construction of the new craft to North American Rockwell Corporation in 1972. At 122 feet long and 57 feet high, boasting a wingspan of 78 feet and weighing in at 165,000 pounds, the orbiter was a mammoth machine. It was also our first reusable orbital spacecraft, since the orbiter could be refurbished and relaunched, and the shuttle's two solid-rocket boosters, which it shed before leaving the atmosphere, were also salvageable and—with a fair amount of effort and expense—reusable. The only component that was expendable was the system's giant external liquid fuel tank. The orbiter was designed so that it could be flown remotely, though it never actually was, and plastered with over twenty-one thousand ceramic thermal-protection tiles that allowed it to descend belly-first into the atmosphere. Powered by three onboard main engines and its two boosters, the shuttle generated seven million pounds of thrust at launch. Five onboard computers provided the brains for an electrical system that involved some 600 harnesses, 7,000 connectors, 120,000 wire segments, and 228 miles of wire.

It was, as they say, *complicated.*

Looking for a Hit

Construction of the shuttle orbiters commenced in the midseventies. The first, named *Enterprise*, was unveiled in 1976 and used for a series of approach and landing tests starting in August of 1977. *Enterprise* was originally going to be called *Constitution*, 1976 being the nation's bicentennial and all. This was a nice thought, but historically inept. While the nation declared its independence in 1776, the Constitution wasn't ratified until 1789. Thus, a more appropriate name for the vessel would have been *Declaration*, or possibly *The Articles of Confederation*, which admittedly lacks panache. All such political considerations fell by the wayside, though, when determined fans of the television series *Star Trek* mounted a successful write-in campaign to have the vessel named after the *Enterprise*, their favorite Federation starship, instead.

Enterprise never flew in space. It was made only for testing. The first shuttle orbiter to be launched was called *Columbia*, and it arrived at Kennedy Space Center in 1979.

America was a different place than it had been ten years earlier, when a kid from small-town Ohio named Armstrong brought home pieces of the moon. It's foolish to try to characterize the mood or emotions of a nation, especially one with 205 million citizens, as the United States had in 1979. The fifties might have been an era of mass conformity in America, but there were plenty of outliers—Cassadys and Kerouacs, furtive commies and fluoride conspiracists—to be found lurking in public libraries and late-night diners across the continent. The late sixties might have had an efflorescence of hippies and flower children, but it also had a thriving middle-class culture open to the pious blandishments of Richard Nixon and Hubert Humphrey.

Still, the United States was a dispirited nation in the late seventies. President Jimmy Carter famously spoke of a "crisis of confidence" in a speech he gave in July of 1979. The country had watched Saigon fall to communism on color T V, pointing up all over again how we'd lost tens of thousands of lives and billions of dollars in trying to prevent what now seemed inevitable. We'd waited in long lines for gas and suffered through something called stagflation. *Skylab* fell out of the sky. In November of that year, student supporters of the new Islamic government of Iran took fifty-two American embassy staffers hostage in Tehran and seemed unbothered by any obligation to give them back. A shambolic American military rescue operation called Eagle Claw failed miserably, leaving eight American servicemen dead in the Iranian desert without ever having fired a shot. Emboldened by American impotence, the Soviet Union meanwhile invaded Afghanistan. Far from being the world's policeman, as it once appeared, the United States now seemed to some like a senior citizen, feeble and adrift. Voters in the presidential election of 1980 decided that a change was in order, opting in overwhelming numbers for the smiling, optimistic Ronald Reagan over schoolmarmish Jimmy Carter. America wasn't failing, Reagan promised. It was the "city on a hill," the chosen land—and it was time to start acting like it again.

NASA, too, was a different place at the end of the seventies than it had been a decade earlier. In the wake of congressional abandonment of crewed space exploration halfway through the Apollo program, the agency began to push its undertakings not as necessary responses to the Soviets or even as wonders of science and curiosity in their own right. Rather, NASA tried to conform to contemporary mores. In the seventies the shuttle was touted as being reusable and thus in tune with the decade's ecological concerns. In the Reagan era, by contrast, the shuttle was peddled not as a triumph of recycling but as a cosmic pickup truck, meant to allow for transport, deployment, and repair of satellites and space habitations as cheaply and efficiently as possible. America was all about business again. NASA had to don its button-down shirts and demonstrate how practical it was. The agency touted "getaway specials" on the shuttle, borrowing a popular advertising phrase to describe opportunities for private interests to pay just a few thousand dollars to have their experiments carried into space. The agency doubled down on its lists of spin-offs, down-to-earth products derived from or created for use in space, like memory foam, scratch-resistant eyeglass lenses, cordless vacuum cleaners, and cochlear implant hearing aids. Shuttle astronauts called themselves the "Ace Satellite Repair Company" and held up FOR SALE signs upon recovering errant communications hardware.

First Flight of the Space Plane

The first space shuttle was launched from Kennedy Space Center on 12 April 1981, exactly twenty years after Yuri Gagarin became the first person in space. *Columbia* was commanded by the gifted and taciturn Apollo moonwalker John Young and piloted by his rookie partner, native Texan Bob Crippen. Despite his inexperience, Crippen seemed up for the challenge—a big change, given that his initial thought on seeing the shuttle orbiter was that "we've screwed up bad, this is never going to work." Apparently, he'd changed his mind. But not everyone had. It was the first time in NASA's history that the functionality of a spacecraft would be demonstrated by means of a crewed orbital flight. In essence, the mission was a test of a huge, extremely complicated machine, only this time with two human lives on the line. John Young's wife, Susy, didn't expect him to survive. Young himself estimated his chance of death at 50 percent.

Initially, the orbiter, its external fuel tank, and its two solid rocket boosters were all painted a brilliant white. This made for pretty pictures. The shuttle standing on its pad was a vision in ivory, made of marble or constructed of cloud, a sort of technological Taj Mahal. This changed when NASA realized the latex paint on the external tank added some six hundred pounds to the spacecraft. Starting with the shuttle's third launch, the tank was left unpainted, leaving it a sort of bad-tooth brown. But on 12 April, the spaceship wore white. Its maiden flight, designated STS-1, caused considerable drama. Upon opening the orbiter's payload bay, Young and Crippen observed that a number of thermal tiles were missing from the pods at the aft end of the orbiter that contained the shuttle's maneuvering engines. This was, of course, problematic, as the tiles were the orbiter's primary protection against the extreme heat of reentry. While a few missing tiles on the pods weren't critical, their loss suggested that there might be problems elsewhere—like on the vulnerable underside of the craft. NASA flinched. This was exactly the scenario the agency had hoped to avoid. Its fears mounted to the point where NASA asked the National Reconnaissance Office to use one of its spy satellites to photograph the shuttle's underside. As chronicled in Rowland White's 2016 book *Into the Black*, this was a complicated operation, as the NRO was still an agency that didn't exist—officially, anyway—and any hint of its involvement was strictly forbidden. Astronaut and former NASA administrator Dick Truly described the machinations involved in working with the spy center as a story of "how two entire national space cultures meshed with each other, and sometimes collided." In the end, though, the NRO was able to provide images indicating that there was no significant tile loss in the shuttle's most critical area, so the calculator crew at mission control gritted their teeth and proceeded with the flight plan. There wasn't much the ground could do at this stage of the mission anyway. After thirty-six orbits, the astronauts prepared for their fiery plunge into Earth's atmosphere and a landing at Edwards Air Force Base in California.

Given the risks in life and investment and the possible pyrotechnics involved, it wasn't surprising that a sizable portion of the world's population tuned in on television to watch, expecting disaster, as Young and Crippen brought *Columbia* out of the black and into the blue. Veteran flight director Gene Kranz expressed certain observers' ambivalence about the new vehicle well. Kranz was a big man, intimidating and blunt, with craggy *On the Waterfront* features and a reputation for fearsome dedication to duty. Speaking about the first shuttle flight with an interviewer for the television series *When We Left Earth*, he opined that the shuttle was the greatest example of American aerospace engineering ever built. On the other hand, he also said that he "prayed a lot" during that first flight. Lots of people did. And maybe it worked. After traveling for three days and over a million miles in space, the shuttle was twenty-seven seconds ahead of schedule when it appeared in the bright sky over Edwards. Loudspeakers broadcast the radio transmissions between the astronauts and mission control to the 250,000 spectators on hand. Flags flew, music played, and people cheered as *Columbia* touched down in the desert. "NEW ERA USHERED IN BY SHUTTLE," wrote the *Chicago Tribune*. "It was one fantastic mission," John Young commented. "*Columbia* is phenomenal, an incredibly amazing piece of machinery. I think the American public is going to get its money out of this baby." NASA was back.

And NASA, it turned out, was dreaming.

The First Disaster

The initial flights of the space shuttle included a string of successes. In 1982 STS-5 deployed the first of many military and commercial satellites. Physicist Sally Ride became the first American woman in space on STS-7 in 1983. Former air force fighter pilot Guion Bluford entered the history books as the first African American in space on STS-8 later that year, and NASA flew its first European Space Agency astronaut, West Germany's Ulf Merbold, on STS-9.

In February of 1984 Bruce McCandless II and Bob Stewart made history's first untethered space walks on the tenth shuttle mission (numbered STS-41B, due to a numbering convention too arcane and pointless to be explored here), when they each left the safety of *Challenger* and flew the manned maneuvering unit, the "Buck Rogers jetpack," a football field's distance away into space. The jetpack was a nifty machine that was used on two subsequent missions to rescue three derelict satellites. It was also somewhat risky, which is why it became an icon of the shuttle era—or at least, *part* of the shuttle era. These were the We Can Do Anything years, the height of NASA's ambitions, a time when the agency was driving itself hard to prove the worth of its Swiss Army Knife of a space vehicle. It was willing to do whatever it took to bring more business to the shuttle. In a 1982 memo to the astronaut office, administrator Ray Dell'Osso reported back on talks he'd had with top NASA brass on upcoming operations. Shuttle chief general James Abrahamson, he said, wanted to "exploit the [shuttle] to its fullest extent and dramatically change the way we [NASA] do business and revolutionize our thinking in this arena. . . . Abrahamson wants to make EVA [i.e., space walks] a routine way of doing business to give the user community confidence in our capability. . . . Also, as part of his sales pitch, he wants to put a 'selling' team together to go out into industry and tell them what we plan to do."

Also on STS-41B was mission specialist Ron McNair. An African American member of NASA's groundbreaking astronaut group eight, McNair was born in Lake City, South Carolina in what the calendar said was 1950. He eventually found out, though, that not much had changed in Lake City since the previous century. A bibliophile, nine-year-old Ron went to his local library to check out a book—a math book, specifically. There he was told that the library didn't lend to Black people, as the notion of an African American doing algebra was evidently too much for the citizens of Lake City to bear. When Ron insisted that he be allowed to check out his book, librarians called the police. A dust-up ensued, and Ron emerged triumphant—and with plenty of ciphering to do. Despite such institutional obstacles, McNair never gave up his love of learning. It led him eventually to North Carolina AT&T and then to the Massachusetts Institute of Technology, where he earned a PhD in physics. He worked for a time at the Hughes Research Laboratory in California, specializing in the physics of lasers, before he joined the astronaut corps as a mission specialist in 1978. He flew on STS-41B, the "jetpack mission," in 1984, and was aboard the ill-fated *Challenger* again when it disintegrated over the Atlantic Ocean shortly after takeoff in January of 1986. McNair was a polymath, a physicist with a black belt in karate and a love of playing the saxophone. Maybe more importantly, he was relentlessly upbeat, confident but self-deprecating, quick with a smile or quip. His death robbed NASA of not only one of its most gifted astronauts but also one of the best loved.

As shuttle flights became more or less routine, NASA began looking for ways to democratize the space mission experience. In April 1985 U.S. senator Jake Garn of Utah flew on *Discovery* as a payload specialist for STS-51D. Garn's jaunt wasn't popular in the astronaut corps, where a number of would-be rocket riders had been patiently waiting for a seat on the shuttle. It didn't go over so well with the maintenance crew, either, as the senator spent a substantial amount of the flight throwing up. But at least Garn was a former naval aviator. NASA next signaled that it was interested in putting civilians of all stripes—experienced or not—in the shuttle. Veteran newsman Walter Cronkite applied for a spot. Singer-songwriter John Denver was interested, too, but ultimately President Reagan determined that a teacher should be the first nonpolitical civilian to get a place on the space plane.

It was a smart decision. The agency's well-publicized search for an educator to send to the stars found exactly the right person. A spunky, unpretentious, thirty-seven year old from New Hampshire, Christa McAuliffe was lauded by friends, neighbors, and students as relentlessly positive, curious, and energetic. Like Ellen Ripley in the popular sci-fi flick *Alien*, even down to her shoulder-length perm, McAuliffe was feminine but sturdy, a mom and career woman with a determined set to her jaw when she wasn't smiling, as if she could handle a pulse rifle as well as a paring knife. We knew she was nervous. Anyone would have been. But she made it clear she'd be fine. She was confident in the skills of her commander, Dick Scobee, and her crewmates— pilot Mike Smith, mission specialists Ron McNair, Judy Resnik, and Ellison Onizuka, and payload specialist Gregory Jarvis. She believed she could rely on the expertise of NASA management.

And so, on 28 January 1986, the day President Reagan was scheduled to give his fifth State of the Union address, *Challenger* swam up off Launch Pad 39-B into a brilliant blue sky. The big dart flew for a little over a minute. That's how long it took for malfunctioning O-rings to allow superhot gases from one of the solid rocket boosters to cut a hole in the orbiter's main fuel tank. The shuttle broke apart at an elevation of forty-six thousand feet, approximately nine miles up, in a chaotic tangle of divergent vapor trails. It was an image unlike any we'd ever seen before, an irrational snarl that is nevertheless instantly recognizable today to anyone who saw the unthinkable happen. Parts of the orbiter, including the crew cabin, continued to ascend for perhaps twenty-five seconds afterward, even as the shuttle's rocket boosters snaked away, bereft of guidance from the shuttle's computers. Inexorably, the cabin's ascent slowed.

And then the astronauts fell.

Parts of *Challenger* are still being found. A lot of what was located in the weeks after the accident, including the crew cabin, came to rest on the floor of the Atlantic Ocean, a hundred feet below the surface. Suddenly, the question of who was going to fly on which mission became moot. President Reagan appointed what came to be known as the Rogers Commission, an investigative panel that included such talented individuals as Neil Armstrong, Sally Ride, and the brilliant physicist Richard Feynman. The commission concluded its review of the disaster with a searing indictment of institutional arrogance at the nation's space agency. NASA, said the committee, had information indicating that the frigid temperatures on launch day might compromise the integrity of the O-rings but decided to launch anyway. It was a decision directly responsible for the loss of seven lives.

There had been space-related fatalities before. The astronauts of *Apollo 1*, of course. Vladimir Komarov on *Soyuz 1*. The three-man space station crew of *Soyuz 11*. But the most recent of these accidents had occurred fifteen years earlier, and none had been broadcast on network television on a day when NASA felt so convinced of its own excellence that it ignored clear warnings about the dangers of a cold-weather launch.

There was plenty of blame to go around. Maybe NASA had simply forgotten it could fail. Or maybe the problem was deeper than that. Some said that the shuttle was bound to fail because it was engineered to do too much for too many, which made it too large and too complicated to be truly safe. Max Faget remarked not long after the first shuttle flew that the spacecraft was already obsolete, the victim of bureaucratic delays that made its technology old even before it was used. Years later, one NASA administrator remarked that "Jesus Christ himself could not 'fix' the shuttle. It was a compromise design from the first." Whatever the reason, NASA suffered not only a gut-wrenching loss of seven good people but also a huge hit to its credibility.

"Early 1986, boom," said astronaut Joe Kerwin, looking back. "Here comes the *Challenger* accident, and it just turns NASA upside down. Never before that had anybody thought that NASA was less than the best-managed federal agency ever. Immediately after, we found that we were the worst federal agency ever. You know, the tables just turned around completely."

A Change in Plans

The shuttle program changed significantly after the *Challenger* disaster. It took two and a half years before another American orbiter left Earth. During that time NASA adopted a culture of caution. The Lost Gonzo days of the early shuttle—the days of Buck Rogers jet packs and rampant Go Fever—were over. In fact, Washington decided that the agency would be getting out of the commercial satellite deployment business altogether. The shuttle entered into its second distinct operational phase, the We Just Work Here years, marked by significantly increased safety procedures and a determination to stick to the meat-and-potatoes business of space transportation. While former shuttle chief general James Abrahamson had pushed astronaut EVAS as a way of demonstrating NASA's satellite launch and servicing capabilities, the agency now frowned upon such activities. A 29 October 1988 memo from NASA manager Donald R. Puddy stated that "in no event should EVA be encouraged unless it is clearly required and in the best interests of the program."

Meanwhile, the U.S. Air Force, a major shuttle customer, decided to reduce its reliance on the space plane in favor of its own Titan rockets, pursuing what it called a "robust mix of [satellite] launch vehicles for assured space access." Ironically, these vehicles, too, were plagued by failures in the late eighties and early nineties; in 1993, for example, a Titan IV carrying a classified payload blew up two minutes after takeoff from California's Vandenberg Air Force Base. Though the shuttle did carry some additional military payloads after 1986, NASA and the air force largely went their separate ways after the *Challenger* disaster.

Hubble

The agency's "Return to Space" flight, STS-26, took off on 29 September 1988 carrying a five-person crew, all veterans of spaceflight, into low-Earth orbit. Over the next fifteen years, the shuttle logged a total of eighty-one flights, performing such varied tasks as assisting in the construction of the International Space Station, deployment of secret military satellites, and, perhaps most importantly, deploying and later servicing the Hubble Space Telescope.

While it was Hermann Oberth who dreamed up the idea of sending an observatory into space, American astronomer Lyman Spitzer initiated a serious discussion of the benefits to be derived from such an instrument with a paper he published in 1946. First and foremost among such benefits, he wrote, was the ability of an orbital telescope to make observations unimpeded by the various distortions caused by the atmosphere, our shimmering ghost swamp of gas and dust. The atmosphere blurs visible light, causing stars to twinkle and making it hard to see our neighboring planets and more distant objects. It also hinders or totally absorbs other wavelengths of electromagnetic radiation, making observations in such wavelength ranges as infrared, ultraviolet, gamma ray, and X-ray difficult or virtually impossible.

Spitzer became a tireless advocate for the idea of a space observatory—the Large Space Telescope, it was originally called—and over the years, managed to convince many of his colleagues of the value of such a tool. They, in turn, lobbied their lawmakers to support the vision. Congress was not so easily swayed. But after several fits and starts, construction of the space telescope commenced in 1979 at the Lockheed Corporation's facility in Sunnyvale, California. Originally scheduled for launch in 1983, the mission was repeatedly delayed due to developmental problems. A subsequent 1986 launch date for a Hubble deployment mission commanded by John Young was postponed when *Challenger* fell out of the sky.

Data and Diamonds

An athlete as well as a thinker, Edwin Hubble was a Rhodes scholar, army officer, boxer, and lawyer whose interest in the scientific aspects of astronomy eventually lured him into a career in astrophysics. It was Hubble, working at Pasadena's Mount Wilson Observatory in the 1920s, who demonstrated that nebulae, the cloudlike images visible in the night sky, are in fact galaxies beyond our own Milky Way, each consisting of millions—possibly billions—of stars. Later, using shifts in the light spectra emitted by stars in these nebulae, he laid the groundwork for the discovery that other galaxies are moving away from ours, and that the galaxies farthest away are moving at the greatest relative velocities. The significance of his work made the pipe-smoking, collegial Hubble, a son of the Midwest, a logical choice for commemoration. In 1983 the Large Space Telescope was officially renamed the Hubble Space Telescope.

The Hubble that eventually flew was not quite as grand a machine as its designers originally envisioned. Even so, the final product was a magnificent achievement, what some stargazers have described as the biggest technical improvement in astronomy since Galileo fashioned his spyglass in 1609. Forty-three and a half feet long and fourteen feet wide, the satellite is often likened in size to a school bus. In space, it looks like a giant telephoto lens wrapped in aluminum foil, with the business end of a rubber spatula attached as a lens cover. Its main mirror has a diameter of 2.4 meters. Holland Ford, an astronomer affiliated with the University of Wisconsin and the Space Telescope Science Institute, calculated that Hubble could detect a firefly as far away as the moon. Further, if there were *two* bugs, and they were more than nine feet apart, Hubble could tell if there were *two* fireflies or just one really bright one. One report said the satellite, which ultimately cost $1.5 billion to develop, would be capable of distinguishing the period at the end of a typewritten sentence from a mile away.

The Hubble deployment mission, designated STS-31, launched on 24 April

1990. Statistics related to the mission are fascinating. First of all, Hubble weighed twenty-four thousand pounds. At the time it was the largest payload a shuttle orbiter had ever hauled into space—something akin to stuffing a sofa into the back of a Subaru and then driving it up Pike's Peak. *Discovery* was also required to lug the orbiting observatory some 380 statute miles above Earth, making the mission the highest that an orbiter had ever flown. While this altitude was required to get Hubble clear of the planet's atmospheric distortion and made for some arresting photography, it was also somewhat intimidating. As STS-31 astronaut Kathy Sullivan points out in the documentary series *When We Left Earth, Discovery*'s fuel supply upon reaching its designated orbit was already more than halfway depleted.

Deployment of the telescope was briefly troubled by a balky solar array. A bigger problem arose some weeks later when the instrument's first images were transmitted to Earth. It turned out that Hubble's main mirror had been manufactured incorrectly. The failure was slight, but it was significant enough to throw the project's viability into doubt. It was also a major black eye for NASA—one that persisted until the agency was able to mount a repair operation in 1993. The operation—sometimes described as adding "contact lenses" to the telescope—was a major success, as indicated by the fact that data from Hubble has to date been used to support some eighteen thousand scientific studies and papers.

Hubble was used in finding all five of Pluto's moons. It watched chunks of a disintegrating comet slam into Jupiter (an incident as rare and exciting to planetary astronomers as catching a fight between a sperm whale and a giant squid would be to marine biologists), detected water plumes from Jupiter's moon Europa, and charted the remains of a "kilonova" believed to be responsible for a very rare and powerful event called a gamma ray burst. Hubble is responsible for our current understanding that most galaxies have at their center a massive black hole. The space observatory has seen stars forming, asteroids disintegrating, and moons changing orbit. In 2022 alone Hubble spotted the farthest and oldest star ever found, Earendel (Old English for "morning star"), some 12.9 billion light years away, and was used in 2022 to confirm discovery of the largest *comet* yet observed—an eighty-mile-wide, five-hundred-trillion-ton behemoth called Bernardinelli-Bernstein, which is visiting our orbital neighborhood on what's estimated to be the comet's three-million-year circle around the sun. It's been a wild ride. According to space historian

William Burrows, Hubble has "lavished Earth with knowledge so profound that even its users [have been] stupefied."

Almost as important as the science is the beauty. Data counts, but so do diamonds. Because of Hubble's ability to capture images on the visible light spectrum and the brilliant production work of the technicians who take Hubble's data and translate it into color, depth, and perspective, the satellite has become a public favorite, what astronaut John Grunsfeld called a "science celebrity." The images sent back to Earth astonish even jaded stargazers, much less those of us who wouldn't know dark matter from Darth Vader. Indeed, Hubble's postcards are so bizarre and ethereal that they sometimes seem more like fantasy art than pictures of actual phenomena. The size, shimmer, and sheer spectral weirdness of the images—the odd, hourglass-like Southern Crab Nebula, for example, several thousand light years away—boggle the imagination and make prophets and dreamers of us all. Some of us pay therapists to tell us we're important and unique. Hubble and other space telescopes remind us just how galactically marginal we all are. The truth is somewhere in the middle. We are small creatures on a tiny planet orbiting a sun that's not even situated in the center of our own local star cluster. We live for less than a moment on creation's timeline. We apprehend the world through brains the size and weight of a homemade meatloaf. And yet our minds can imagine the stretch and scope of at least some shadow of infinity, which is the shape of God.

And that brings us to Kalpana Chawla.

Columbia

On 1 February 2003 the space shuttle *Columbia* broke apart as it roared through Earth's atmosphere on its way to a planned landing at Kennedy Space Center. Seven astronauts died in the disaster. Their bodies fell along with shredded pieces of the orbiter across thousands of square miles of east Texas and Louisiana. An intensive search and collection effort ensued, with thousands of Texans joining forces with NASA to comb the forests and fields for pieces of the doomed spacecraft. A sort of rocket forensics investigation confirmed what some had suspected all along. The cause of the catastrophe occurred at launch, when a piece of polyurethane foam insulation broke off from the orbiter's external tank and struck the leading edge of *Columbia*'s left wing. The impact damaged the carbon-reinforced wing to such an extent that on reentry, super-heated atmospheric gases entered and compromised the inter-

THE BUTTERFLY AND THE BULLET | 155

nal wing structure. The wing failed. The orbiter went into a violent spin and disintegrated while traveling at a rate of 11,500 miles per hour, approximately fifteen times the speed of sound.

The *Columbia* disaster never got the attention *Challenger* did. Fewer people cared to acknowledge it. The shuttle was supposed to have been *fixed*. Surely an accident like this couldn't happen again. Much of the American public was like a person who ignores a bad dream when she knows she's supposed to have been cured of her nightmares. And while many Americans remember Christa McAuliffe as the face of the *Challenger* accident, the crew of *Columbia* remains mostly anonymous. For the record, they were Rick Husband, William McCool, Michael Anderson, David Brown, Kalpana Chawla, Laurel Clark, and Ilan Ramon.

Each of the doomed orbiter's astronauts is worthy of commemoration, of course. But the story of mission specialist Chawla is significant for its illustration of the way that the American space program continues to inspire men and women around the world. Her husband, Jean-Pierre Harrison, provides important insights into Chawla's thinking in his 2011 book *The Edge of Time: The Authoritative Biography of Kalpana Chawla*. Though she was female and born in gender-segregated northern India in 1962, Chawla's early life was similar to that of Neil Armstrong and other American astronauts in at least one respect: she was fascinated by airplanes. Born into a Hindu family, she soon found she had no use for religious or political dogma of any sort. Nevertheless, she remained respectful of her culture and was a vegetarian all her life. Her birth name was Montu, but she adopted Kalpana, a Hindi word for "imagination," when she was very young. In 1982 she earned a bachelor's degree in aeronautical engineering from Punjab Engineering College in Chandigarh, the self-proclaimed City of Beauty, a municipality situated on maps of India like a bindi on the nation's forehead. She then made her way to the United States, where she studied at the University of Texas-Arlington and went on to earn a PhD in aeronautical engineering from the University of Colorado-Boulder. She married an American and obtained American citizenship—which required, among other things, forfeiture of her Indian passport. She learned to fly. And she and her husband fell in love with the West—mountains, deserts, hoodoos, and half domes. Chawla dubbed one valley she hiked, a green pocket filled with flowers, as the "heaven place."

After joining NASA in 1994 as a member of the fifteenth astronaut class, the Flying Escargots, Chawla became the first woman of Indian descent to enter space. In doing so she became an exemplar not only for Western females but indeed for women around the world. She was a diminutive figure, generously listed on NASA's website at five feet, four inches, with a shock of jet-black hair and a scientist's skeptical smile that broadened occasionally into a little girl's bashful grin. On her first spaceflight, STS-87, she made the comment, "You are just your intelligence." Agree or disagree, it's a fascinating if somewhat cryptic statement, completely unlike the rhetorical wallpaper— "Work hard"; "It's neat up there"—favored by many NASA rocket riders. She was a private person reluctant to discuss her inner life. What *is* known is that she uprooted herself as a college student to travel eight thousand miles to the United States, where she earned a chance to do what millions of young people dream about doing. In accordance with her love for the American West, Chawla's cremated remains were scattered in Utah's Zion National Park.

The accident investigation and suspension of flights after the *Columbia* crash lasted for over two years and seemed like a doleful déjà vu. When NASA finally resumed operations, the shuttle entered into its third and final operational stage, the Ivory Albatross era. Discussions began regarding how and when to retire the spacecraft and whether its replacement should be a newer, bigger version of the winged vessel or something completely different, like a rocket-borne capsule. Shuttle missions were largely limited to missions to the International Space Station, where astronauts helped to construct the new habitation and a Russian Soyuz spaceship was kept parked at the ISS for the shuttle crew to use if the orbiter was damaged or malfunctioning. It was a tacit admission that the shuttle was a flawed space delivery system, to be treated warily. A tacit admission—and a tardy one.

Denouement

Even act 3 of the shuttle era had some bright spots.

Eileen Collins was born in Elmira, New York, in 1956. She attended Catholic schools in town and still remembers the "duck and cover" drills she and her classmates performed to prepare for a nuclear attack. Her childhood was shadowed by her father's alcoholism, a condition that got so bad that Collins's mother eventually threw him out of the house. When her mother was briefly institutionalized for mental-health issues, young Eileen found herself in charge of herself and her two younger siblings. By her own account, she was a so-so student in high school, possessed of no particular talents or ambitions. In fact, it was her realization of this fact that led her to a sort of epiphany during her graduation ceremony, when she figured out she was going to have to outwork others in order to make something of her life. She enrolled in the air force ROTC program at Syracuse University and—partly through sheer force of will—excelled in both her mathematics course work and the military studies the ROTC program required.

The air force, the reader will recall, was the service in which officers jumped out of high-altitude balloons and rode rocket sleds till their eyes bled at the dawn of the space age. It was not an organization noted for a particular interest in equity or inclusion. In the early sixties, for example, African American air force pilot Ed Dwight was a good bet to join NASA's astronaut corps. He was, that is, until he entered test pilot school and ran into the air force's legendary Chuck Yeager, who—according to Dwight, at least—acted on the basis of racial animus to torpedo his prospects for a spaceflight. However, and to its credit, the air force was in the seventies beginning to look for women to fly its jet airplanes. Collins managed to penetrate the leather curtain to become one of the service's first female pilots since the days of World War II, when distaff fliers like Jackie Cochran and Nancy Love routinely ferried even the biggest bombers across the continent. After graduating from Syracuse and earning a commission, she began working her way through various assignments, including command of a C-141 Starlifter transport plane—a *whale* of an aircraft, some 168 feet long—that airlifted American medical students out of Grenada in 1983 during the course of a violent civil conflict on the island.

Collins eventually qualified for USAF test pilot school, traditional home of the airborne alpha male, and was working in that capacity when she was selected for the astronaut corps in 1990. Collins shows no outward trappings of the razor-sharp pilot she is, and she seems never to have made an enemy. Gracious and good humored, she comes across in person like everyone's favorite fifth grade teacher—until she makes a suggestion or comment, which somehow carries the weight of command, and people around her suddenly find themselves tucking in their shirts and stiffening their spines. She became the first female shuttle pilot on STS-63 in 1995. Later, she was the first woman to command a space shuttle mission with STS-93, the Chandra X-Ray Observatory deployment flight, in 1999—a mission that almost ended in disaster shortly after takeoff due to a hardware malfunction. Collins also commanded

STS-114, which was the first shuttle to fly after the *Columbia* disaster in 2003. She was inducted into the U.S. Astronaut Hall of Fame in 2013.

Epitaph

The space shuttle never became the dream ship it was advertised to be. It was far more expensive than initially envisioned. It had no viable escape system for its astronauts. It was also far less reusable than initially claimed, and it was plagued by safety problems that killed fourteen men and women on two missions. Still, it was a major phase of American space exploration. The five shuttle orbiters to leave the planet—*Columbia, Challenger, Discovery, Atlantis,* and *Endeavour*—flew 135 missions from 1981 through 2011, covering 542 million miles and 21,152 orbits of Earth. The vessels deployed 180 satellites, launched the ambitious interplanetary probes *Magellan* and *Galileo,* docked nine times with the Soviet (later Russian) space station *Mir,* and helped construct the International Space Station. The Russians liked the shuttles so much that they built one for themselves, though it flew only once.

Though never quite as startling in its design or bold in its execution as the Apollo hardware, the space shuttle had its own brand of beauty. It was a faulty future. It was an elegant killer. It was, as astronaut Story Musgrave put it, a butterfly bolted to a bullet.

NASA's Eleven Greatest Missions

1. *Apollo 11*: Nothing beats the first feet on the moon.

2. *Voyager*: Humanity's Hallmark cards are now in interstellar space, courtesy of two mostly mute but still traveling space probes launched in 1977. Citizens of the Sirius solar system, say hello to Mr. Mozart!

3. STS-31 (the Hubble Space Telescope deployment mission): Images from Hubble are like seeing the ocean when the fog finally lifts. The data from the Little Space Telescope That Could may eventually be eclipsed by the work of the James Webb Space Telescope, but it's going to be tough to beat the legacy of this thirty-plus-year workhorse.

4. *Apollo 8*: In 1968, as the CIA warned that the Russians were about to beat us to it, NASA bet big on a possibly premature flight around the moon—and won.

5. Mercury-Atlas 6: John Glenn, Everybody's All-American, took the controls for our first orbital mission in February of 1962. At last the Soviets could see us coming.

6. Mercury-Redstone 3: The good guys put a man in space, only a few weeks after the Kremlin did it first. The wait seemed endless. Commander Shepard would really like to blast off now, fellas.

7. STS-1: In April of 1981 astronauts John Young and Bob Crippen chanced a white knuckles jaunt on a brand-new vehicle called the space shuttle. Young gave himself a 50/50 chance of survival. His wife thought he was being optimistic.

8. Mars 2020: Okay, so we don't have flying cars yet. But this ongoing exploratory mission to Mars deployed an industrious rover and a nifty autonomous helicopter called *Ingenuity* (or *Ginny*, for short)

that flew dozens of missions before crashing and gave us our best look yet at the desiccated but still fascinating Red Planet.

9. Double-Asteroid Redirection Test (DART): Earth strikes back! In 2022 a NASA probe traveled seven million miles into space to impact and alter the orbit of an asteroid, proving that (a) we've got pretty good aim and (b) asteroids can in fact be pushed around. Someday we're going to be thankful for this little mission and the scientists who are working to prevent a catastrophic deep impact.

10. Skylab 2: The first crewed *Skylab* mission featured incredible feats of engineering and derring-do, though most of us preferred to watch *The Mary Tyler Moore Show* instead.

11. *Viking 1* and *2*: The first probes to "soft land" on Mars performed admirably and sent back great images but failed to find evidence of life. Or did they? While the mission's experimental results were viewed at the time as a disappointment, the data collected by *Viking* are again a hot topic of discussion among exobiologists.

14

Sleeping with the Russians

A space station sub-rivalry between the United States and Russia eventually resulted in creation of the International Space Station, an experiment in cooperation that has now lasted a quarter of a century.

After it failed to develop a rocket that could put its cosmonauts on the moon, the Soviet Union shifted focus. It wasn't interested in rock collecting on the lunar surface after all. *Detskaya igra*, Moscow grumbled, in the tones of one who's no longer enthused about a game he can't win. *Child's play.* The real challenge in space, the Kremlin now said, involved creation of human habitats suitable for long-term scientific study and experimentation.

Accordingly, the Soviets put the first space station in orbit on 19 April 1971, two years before *Skylab* was launched. Called *Salyut* ("salute," in English, a tribute to Yuri Gagarin's spaceflight ten years earlier), the station was a minor marvel—small, around twenty meters in length and four meters in diameter, with three pressurized compartments. Three cosmonauts visited the station on *Soyuz 11* in June of 1971 and stayed for a total of twenty-three days, a new space endurance record. However, the cosmonauts died on their return to Earth when their Soyuz capsule depressurized before reentry. Depressurization means, among other things, a rapid loss of oxygen. The cosmonauts suffocated quickly. To this day, Victor Patsayev, Vladislav Volkov, and Georgy Dobrovolski remain the only human beings to have died in space.

By contrast with America's Skylab project, which involved one station and had a relatively brief run, from May of 1973 through November of the following year, the Salyut program involved the launch of *eight* space stations, six of which actually achieved orbit and hosted cosmonauts. Of this number, four were civilian habitations, while two others (dubbed the Almaz, or "diamond" stations) were essentially military observation platforms. *Salyut 6* and *7* were the most successful of the Soviet stations, with by far the greatest number of occupancy days. *Salyut 3*, an Almaz platform, is remembered chiefly because it earned the dubious distinction of being the first orbital vessel on which military weaponry (a 23 mm rapid-fire cannon) was test-fired. At one point essentially abandoned and then subsequently—and somewhat miraculously—resurrected by cosmonauts in 1985, *Salyut 7* was finally allowed to burn up in Earth's atmosphere in 1991, nine years after its launch.

Just a Little *Mir* Cooperation

The next-generation Soviet space station program was called *Mir*, or "peace." It was active from 1986 to 2001 and notable for its innovative module-based construction, meaning that it was assembled unit by unit, with a "base block," or core module, deployed first, and then various other blocks—science laboratory modules, a docking module—added later. In all, the station grew to include seven pressurized modules and several unpressurized components. In this respect *Mir* was a model for its much larger successor, the International Space Station, or ISS, which has been operational for over twenty years now and has something like *seventeen* pressurized modules in use.

Mir's power was provided by several solar arrays attached directly to the modules, four of which arrays were arranged in such a way as to suggest the wings of a dragonfly. The station was widely derided in the nineties for its technological failings. It was called dangerous, dirty, and unreliable. By the time *Mir* returned to Earth in March of 2001, though, deorbiting over the Pacific Ocean, the station had more than tripled its projected five-year lifespan. Before that, it hosted the first period of extended—and occasionally uneasy—cohabitation between the United States and Russia in space. And therein lies a tale.

In the early 1980s President Ronald Reagan announced his support for an American space station that would outshine *Mir* in much the same way that *Skylab* one-upped *Salyut*. The need for a permanent station was much discussed at the time. The science fiction writer Isaac Asimov, a prolific fantasist and colorless prose stylist who could make even the most exciting ideas seem like an early dinner with the in-laws, laid out the advantages of a permanent station in a July 1986 article in *Popular Mechanics*. One advantage, he opined, would be that "space settlements would offer an ideal inducement for space travel. At their distance from Earth, the escape velocity would be very low. Between that and the omnipresent vacuum of space, fuel requirements would be moderate, and advanced methods of propulsion (ion drive, solar wind sailing) might be made practical."

As Bryan Burrough writes in his book *Dragonfly: NASA and the Crisis aboard Mir*, the new American space station, dubbed *Freedom*, was envisioned in 1986 as being "an eight-man station . . . with nineteen scientific instruments, a 'garage' for repairing satellites, four separate laboratories, and a hangar for building spaceships to fly to Mars." Years of redesigns and budget wrangling followed, with the station's projected price rising from $8 billion to $120 billion by 1990. Even the station's supporters were frustrated. "We've spent $4 billion so far and there isn't a nut or bolt to show for it," said one scientist.

By the time Bill Clinton was elected president in 1992, many observers expected the wildly expensive project would be scrapped altogether. Only the frantic efforts of NASA administrator Daniel Goldin to present less costly, more palatable alternatives to *Freedom* saved the idea. An unexpected overture from the Russian space agency helped to seal the deal. As U.S. relations with Moscow improved, Washington opted to work with the Russians to construct what we now know as the ISS.

Thus began a strange chapter in U.S.-Russian relations. There were many causes of the breakup of the Soviet Union, but one of the chief reasons was economic; the USSR's socialist economy couldn't compete with the prosperity of the West, and its failures became increasingly obvious. Westerners watching the Russia that emerged from the wreckage of the Soviet Union in the early nineties hoped that the new nation would be open, capitalistic, and democratic. These hopes were never fully realized, but both George H. W. Bush and Bill Clinton saw Russia in those early days as a promising partner for space operations. The theory seemed to be, *If they can't beat you, join 'em*. Politicians weren't the only ones intrigued by the prospect. As early as 1988, astronauts at Johnson Space Center were asked to propose ways to work with Russia. One shuttle veteran proposed combining work on the fledgling Global Positioning System with the Soviet equivalent, GLONASS, to produce a "joint navigation system," with the navigation channel and orbit management turned over to some trustworthy third party who would guarantee the system's functioning in case hostilities involving one or both of the countries broke out. This idea went nowhere, but clearly the possibility of working with the Russians rather than against them was in the air.

It wasn't just wishful thinking. There were strong practical considerations involved as well. Embracing Russia and its space program was seen as a way to acquire Russian technology, especially its rocket technology. Partnership with

NASA would provide Russians with a positive example of cooperation with the West, and it would keep Moscow's engineers and technicians employed at home rather than working for regimes like those of China, Iran, and North Korea. Infusions of capital were also seen as important in easing pressures on Russia to sell missiles and guidance systems to such nations. In fact, the money was crucial. The Russian space agency, Roscosmos, was so strapped for cash that it had its cosmonauts filming commercials in space for MTV, Pepsi, and Hewlett-Packard computers. And finally, a partnership with Russia would pave the way for cooperation on the construction of a different, and much larger, space station, and give the space shuttle someplace to shuttle to and from.

The Clinton administration saw working with Russia on *Mir* as phase one of a three-part plan that would culminate in the joint construction and operation of the ISS. So it was that the United States agreed in 1993 to pay Russia $400 million for American astronauts to live and train on *Mir*. It was a lot of money, and the payout was characterized by some as "welfare" for the proud but penniless Russian space program. Ultimately, seven Americans spent time on the metal dragonfly, but by the time they started, in 1995, the station was already well past its best-by date. What followed was a largely forgotten chapter in American space history, a three-year sleepover with the cosmonauts in a rickety celestial flophouse. The nations also agreed to let spacefarers from each country travel in the other's vehicles. Thus, cosmonaut Sergei Krikalev became the first Russian cosmonaut to fly on the shuttle in 1994, and astronaut Norm Thagard was the first American to visit *Mir* when he journeyed to the station on a Russian Soyuz spaceship in 1995. A Russian space launch involves a number of pre-launch traditions that cosmonauts, including visiting cosmonauts like Thagard, must endure: a visit to the burial place of Yuri Gagarin, a tree-planting ceremony, and a blessing by a Russian Orthodox priest. On the way to the launchpad, cosmonauts would get out of the bus carrying them to urinate on one of the vehicle's rear wheels—an act said to have begun with Gagarin and thus essential to carry on in the name of good luck. Female spacefarers had the option of peeing before the bus trip and tossing a container of their urine on the wheel.

And that was just the fun stuff. Once on board *Mir*, American astronauts had to endure an onboard fire, an orbital collision with a supply ship, power outages, spotty communications, and a lack of support by both Russian and American ground control teams. If *Skylab* was a drafty Quonset hut, *Mir* was the belly of a snakelike beast, cramped, fetid, leaky, and dyspeptic. "*Mir* is breaking about as fast as they can fix it," said John Pike, a space expert at the Federation of American Scientists, in 1995. "There is so much crud in the plumbing system that it will limit the life of any new equipment they bring up to fix it."

Our astronauts were supposed to be conducting scientific experiments, which some did with more success than others. But all the Americans who visited *Mir* were challenged. Things broke. Machinery malfunctioned, and globules of antifreeze floated in the modules like tiny jellyfish in a cosmic sea. Following orders passed down from Star City, Norm Thagard confined himself to consumption of only officially sanctioned Russian foodstuffs, which he disliked so much that he lost fourteen pounds. As set out in *Dragonfly*, Roscosmos doctors were appalled to learn that Thagard wasn't munching on any of the salty or sweet snack foods cosmonauts enjoyed "off the books." They ordered him to indulge. "You're free to eat anything except your crewmates," they told him. It was unclear whether the Russians had received the same instruction. Thagard's successor on the station, Shannon Lucid, seems to have managed her stay with few problems, though she was simultaneously impressed and alarmed by the amount of improvisation required of her Russian comrades. NASA's astronauts are by contrast extensively trained—in some cases possibly *over*-trained—for every procedure they are scheduled to undertake. Astronaut Jerry Linenger lived on *Mir* for 132 days in 1997, a singularly troublesome year for the station, and returned to Earth with dire warnings about the imminent failure of the aging Russian hardware.

The International Space Station

The U.S.-Russian *Mir* experiment broke no technological barriers, and its scientific achievements were modest. It was, however, an important trial run for cooperation between the two nations in building and operating the far more illustrious ISS. Commenced in 1998 and mostly finished in 2011, though additional modules were added on as recently as 2021, the ISS is a spindly but massive complex of modules transported to space by American and Russian spacecraft and bolted onto a central hub piecemeal over the years. The resulting structure resembles a cross between an industrial pipeline and those "hedgehog" obstacles the Germans erected on the beaches of Normandy before D-Day.

First and foremost, the ISS is a science platform. And as with *Skylab*, a big part of the science relates to the human body and its ability to adapt to weightlessness and increased radiation exposure. But visitors do other sorts of experiments too. According to Space.com, "As the only microgravity laboratory in existence, the ISS has facilitated more than 3,600 researchers to conduct more than 2,500 experiments to date." Among the highlights are the year-long physiological study of the effects of space travel on astronaut Scott Kelly, as compared to the general health of his earthbound identical twin, former astronaut Mark Kelly; creation of a very rare state of matter called a Bose-Einstein condensate; insights into the hardiness of earthly bacteria in space; and the cultivation of vegetables—cabbage, kale, lettuce—for consumption by the astronauts. The ISS succeeded *Mir* as the largest artificial object in space. Roughly the size of a football field (with end zones included, says NASA), it is regularly visible to the naked eye from Earth's surface. It has been continuously occupied since its first modules became operational in 2001 and has hosted 250 astronauts from nineteen nations. The European, Canadian, and Japanese space agencies have also invested time, money, and hardware in the ISS and thus share astronaut flight time. One nation that has *not* visited the station is China, which has been blackballed for years by the United States because of American suspicions of Chinese spying and theft of U.S. technology.

Rambling and relatively spacious, the ISS has been the site of numerous American space endurance records, including those of Peggy Whitson (675 cumulative days in space) and Frank Rubio (371 *consecutive* days in space). The ISS maintains an average altitude of 250 miles by means of re-boost maneuvers using the engines of the Zvezda Service Module or visiting spacecraft. It travels at some 17,500 miles per hour relative to Earth and circles the planet in roughly ninety-three minutes, completing 15.5 orbits per day. It's frequently visible in the night sky, and NASA operates a website that provides up-to-date information on how to spot it. The station has functioned with relatively little fanfare or controversy. Indeed, it is not uncommon to encounter people who have forgotten that it even exists and that astronauts have been orbiting Earth on the ISS for over twenty years.

Astronauts on the ISS are fairly frequently warned about incoming space debris and have become used to dealing with the headache that is orbital space junk. But the biggest threat to the orbital platform's continued viability may well have come in February of 2022, when Russian military forces invaded Ukraine. The United States has heavily funded Ukrainian resistance since the start of the conflict, which has soured the always fragile relationship between Washington and Moscow. Russian space czar Dmitry Rogozin issued a number of inflammatory and hostile declarations about the future of the ISS in the early months of the war, at one point threatening to leave American astronaut Mark Vande Hei stranded in space. Such statements weren't particularly helpful, of course, and the Russian government eventually relieved Rogozin of his duties. Though the fighting in Ukraine continues, joint U.S.-Russian crews have apparently been able to avoid discussing politics on the station, and the ISS is still operational.

Citing maintenance expenses associated with the aging infrastructure of the station, NASA currently plans to let the ISS "de-orbit" (i.e., fall back to Earth) in 2031. The agency is hoping one or more commercial entities will step up to build a successor station, which NASA will help launch, deploy and, eventually, rent space on. A number of companies—SpaceX, Blue Origin, Vast, and Voyager, to name a few—are actively working on the project. Meanwhile, China operates its own space station, called *Tiangong*, or "sky palace." Russia, too, has announced plans to build another space station— this one apparently without help from or cooperation with the United States. Unless some American or Western company is willing to step up, we may at some point in the future have to rely on the Chinese or Russians for "space station time" in the same way we relied on the Russians for space station transportation until recently.

And that would be awkward, to say the least.

15

False Starts, Missteps, and the Promise of Artemis

For years, NASA's been trying to figure out where to go next.
Artemis is the answer. The current answer, anyway.

While the story of America's space exploration program is generally impressive, our efforts haven't always been successful. NASA has lost seventeen astronauts in space vehicles. We came close to losing others on *Apollo 13* (oxygen tank explosion); the Apollo-Soyuz Test Project (toxic gas exposure); and STS-27 (damaged thermal tiles). We've wrecked rockets, boosters, and Mars probes, and we were disappointed to find that the billion-dollar Hubble Space Telescope was deployed with an incorrectly manufactured main mirror. We were lucky not to kill a cross-section of Australians when *Skylab* came raining down on their heads in 1979. They claimed it fell. NASA explained that no, this was called "de-orbiting," but the clarification didn't seem to help.

Aside from the triumphs and blunders, though, there have been a number of "false starts"—programs or projects that were killed, co-opted, or abandoned over the years without ever having a chance to either succeed or fail. Not every idea is worth pursuing. In fact, not every idea is completely serious. An agency can daydream too. Thus, 1967's proposed "manned Venus fly-by," a year-long visit to and return from a planet that can't possibly be used as anything other than a giant pizza oven, was justifiably set out by the curb. EVAR, the agency's robotic space rescue device, an automated lifeguard that looked like a cross between a helpful nun and a haunted refrigerator, fell by the wayside, as did "rescue balls," pressurized pouches for transporting stranded or stricken shuttle astronauts from one spaceship to another. But other now-defunct space plans seem, on second thought, to have been lost opportunities. This is not an attempt to indict NASA. Hindsight is 20/20, as they say, and NASA has always had to manage what it *wants* to do with what Congress gives it enough money to do. Often, promising ideas end up in the trash can simply for lack of cash. Here's a sampler of the agency's wrong turns.

Apollo 18–20

As late as 1970 there were three crewed lunar landings scheduled in addition to the total of six that actually occurred. Three of these planned flights—the original Apollo missions 15, 17, and 20—were canceled, with the remaining missions taking new numerical designations and only going up to 17. Again, the culprit was budget restraints in the aftermath of the heady, spend-anything days of the early sixties. Crews for the lost missions were never formally announced, and spaceflight enthusiasts have spent many hours attempting to project which astronauts would (and would not) have been included on the canceled flights. Additional missions would have meant more lunar samples, more science, and more practice in the notably difficult area of moon landings. On the other hand, there were some even within NASA who felt that three more landings would have produced a lot more risk with only a little more benefit. Best, they opined, to end on a high note. Notably, the Apollo-Soyuz Test Project mission in 1975 is sometimes referred to as *Apollo 18*, as it was launched with Apollo hardware. Also worth noting: cancellation of these missions left NASA with a number of sophisticated, powerful, and expensive Saturn rockets without a lunar-related mission. Fortunately, several were repurposed for use in launching *Skylab* and its crews. Others are now museum pieces, horizontal monuments to sky-high ambitions that ended up going sideways.

NERVA

NERVA is an acronym for "nuclear engine for rocket vehicle application," a serious subject of study for NASA in the late 1950s and 1960s. Touted as a viable method for sending astronauts to Mars, the nuclear propulsion project was canceled in 1971, when ambitious (and expensive) plans for space travel fell victim to budget bashers. But while it lasted, NERVA produced working nuclear thermal rocket engines with impressive performance numbers. By the way, NERVA shouldn't be confused with Project Orion, another sixties-era project that envisioned using a series of controlled nuclear explosions to power spacecraft propulsion. Bizarre as it seems, the method had some solid engineering behind it. In light of current international nuclear testing treaties, however, it would be extremely difficult to prove that such a thing could work, and Project Orion was eventually dropped like a lump of hot plutonium.

Astronaut Bruce McCandless II chalked up the cancellation of NERVA to the following reasons: "high costs for the envisioned missions at the cutting edge of technology, the impact of the Vietnam War on the U.S. economy, the advancement of low-cost robotic technologies for exploration, and the fading of the Soviet challenge to space preeminence. Had the Soviets beaten us to the Moon, 'raising the ante' might well have taken the form of an all-out, probably nuclear-propelled race to 'put a man on Mars and return him safely to Earth by the end of the decade' of the 1970s." Now, fifty years later, we are again confronting the notion that missions to anywhere but low-Earth orbit or the moon will probably need the power of nuclear fission—or, even better, *fusion*. But because the American public was anxious about the risks of nuclear power—un-NERVA'ed, perhaps—at the time, we're not much closer to finding a solution than we were fifty years ago.

The X-33/*VentureStar*

Aerospace engineers have for years been captivated by the prospect of creating a single-stage to orbit (SSTO) reusable vehicle—one that could, in other words, take off from Earth, enter Earth orbit, and return home without the use of additional "stages," or booster rockets. One promising concept in the sixties was the air force's so-called X-20 Dyna-Soar, a plane-like lifting-body precursor of the space shuttle orbiter. But the closest we've come to an SSTO spacecraft came in the 1990s, when Lockheed Martin was awarded a contract to create the X-33. Lockheed Martin's prototype, the VentureStar, incorporated a number of new technologies, including a metallic thermal protection system, composite cryogenic fuel tanks for liquid hydrogen, an aerospike engine, and autonomous (uncrewed) flight control—theoretically possible with the shuttle orbiter but never actually attempted. Funding for the VentureStar was eliminated in 2001 after Lockheed Martin failed to demonstrate the viability of its composite fuel tanks. As that problem may (or may not, depending on whom you ask) have been solved since then, some observers argue that NASA should have kept trying with this next-gen (and presumably much better) space shuttle.

It should be noted that another promising SSTO concept of the era, the McDonnell Douglas DC-X, or Delta Clipper, the first rocket-powered craft to take off and land standing upright, was also snubbed by NASA in the nineties. The vertical landing capabilities of the Delta Clipper, sometimes called the "flying traffic cone," were later adopted and refined by, most prominently, the private space companies SpaceX and Blue Origin. Such landings seemed wildly difficult and impractical until someone started doing them. Now they're commonplace. Whether it was the fault of NASA, Congress, or other political influences, the failure to commit more time and resources to these innovative space vehicles, coupled with NASA's decision to phase out the shuttle program, eventually left us hitching ever-more-expensive rides into space with the Russians on their Soyuz spacecraft. From 2011 until 2020, the United States paid for the privilege of visiting our own space station modules. And it wasn't cheap—over $3 billion over the course of the nine years, by one reckoning. That's a lot of vodka, comrades.

Space Station *Freedom*

In his 1984 State of the Union address, President Ronald Reagan laid out plans for the construction of a gigantic orbital space station, one that would allow Americans (and others) to "follow our dreams to distant stars, living and working in space for peaceful economic and scientific gain." As was the case with another of Reagan's schemes, the Strategic Defense Initiative, or "Star Wars," system, *Freedom* never quite came to fruition. Costs were one concern; technological issues were another. Eventually, space station *Freedom* was discontinued, and elements of its architecture were incorporated into a new plan, championed by President Bill Clinton, to build a space station with Russia. The Soviet Union, America's bitter rival in space for many years, was no more. The nation broke up in 1991, with an apparently less inimical Russian Federation emerging in its place. The Clinton-era plan eventually resulted in the construction of the generally successful ISS, which has been operational for over twenty years as of this writing.

The Constellation Program

NASA's stated agenda of ambition from 2005 to 2009, Constellation was a George W. Bush–era construct that targeted the return of Americans to space beyond Earth orbit in three steps: completing and occupying the ISS, returning to the moon, and landing Americans on Mars. President Obama effectively canceled the program, citing its large projected (and quite possibly understated) costs. Obama's substitute was the more modest Space Launch System—still a thing, by the way—which incorporates the Constellation program's Orion space capsule concept but abandoned the proposed Ares rockets that were to power the Constellation missions. The Trump administration revived the goal of moon and Mars landings and instituted the much-ballyhooed Artemis program, the agency's new lunar landing initiative and hype generator. Artemis aims to make our return to the moon permanent and to use the lunar surface as a test bed and jumping-off point for a future Mars expedition. The Biden administration elected to continue Artemis more or less unaltered, giving NASA some much-needed continuity of purpose. Note that the rocket that will take Orion into the heavens lacks a god name. It's just the SLS—the space launch system. It's not a Thor, a Saturn, a Mercury, or a Nike. It's not even a Vulcan. Those names are already taken. *Zeus* is still available, but perhaps that's presumptuous. Venus might send the wrong vibe. No one wants to shoot into space on top of the goddess of grains and cereals, Ceres, but certainly Athena—goddess of wisdom—would be appropriate. Why the sudden reticence about inspiring nomenclature? It seems to us that NASA is missing an opportunity here.

Arriving at Now

A history of baseball can be told through its players. The history of a war can be told by detailing its battles. A history of the space program is a little more difficult to distill. It's not a matter of studying the astronauts, the most visible actors in this lengthy, uneven film we've made of ourselves. The astronauts don't make policy. They perform the missions they're given. Nor are administrators the best place to look, as these political appointees are factotums of whichever party controls the White House.

Maybe the best way to understand the evolution of the American space program is through politics, international and otherwise. Our early flights were dictated in response to the Cold War and the Soviet Union's head start to the stars. Following the success of Apollo, during a period of political détente and social introspection, we sent up *Skylab* and a number of far-flying interplanetary probes. A renewal of the Cold War in the early eighties brought on a militarization of space policy that lasted through the decade. The demise of the Soviet Union and the emergence of a non-communist Russia in 1990 led to the International Space Station and not a lot else. Successive administrations fumbled the ball of space exploration, going from one ambitious, unfinished project to another. It sometimes seems as if crewed spaceflight—On to Mars!—is trotted out and dangled like a Christmas ornament in front of Congress to attract funding, while NASA secretly hoards its pennies and tries to send up as many serious—but unglamorous—science missions as possible in the meantime.

The Artemis program brings us to the present day—and, in fact, a little beyond. And here's where things get hazy. Looking back is a lot easier than trying to make out what's ahead. It seems probable that robotics, propulsion and spacecraft technology, super telescopes, and artificial intelligence will continue to evolve and will no doubt contribute to our exploration of space in ways that we may not even be able to imagine today. And NASA will have to continue to package its priorities in a way that excites the public and ensures congressional funding. The Artemis program may one day be looked upon as pivotal. It might also be seen as another programmatic misstep, a space station *Freedom*, abandoned in favor of some new idea or ambition. For the time being, though, that's where we're headed: Artemis, and the moon.

A Suitable Deity

The Artemis program's spacecraft consists of an Orion crew capsule wedded to a European Space Agency–constructed service module, all bolted to a massive Boeing-manufactured SLS rocket, the most powerful to leave Earth orbit since the days of von Braun's monstrous moon seekers. (SpaceX's Starship is actually bigger and more powerful, but as of publication date, it has yet to complete a test flight.) After many delays, *Artemis 1* finally launched in November 2022 and notched a more-or-less untroubled twenty-five-day mission to the moon and back, uncrewed but occupied by mannequins and robots. *Artemis 2* is currently scheduled to follow in late 2025, with *Artemis 3*'s crewed lunar landing slated for 2026.

The plan for the post–*Artemis 3* lunar landings is to have the gumdrop-shaped Orion capsule journey to the moon on the strength of its SLS launch vehicle, enter lunar orbit, and dock there with an orbiting space station called "Gateway." There the astronauts will transfer from Orion to a lunar landing vehicle, contracts for construction of which have been awarded to SpaceX and Blue Origin. The astronauts will descend to the moon and take up residence in a base camp, which will be established near the moon's south pole. There, Artemis astronauts will eventually have the services of one or more lunar terrain vehicles to aid them in their exploration and research.

The Artemis program is very consciously named after a female deity, the Greek goddess of the hunt. One of its goals is to land a woman and a person of color on the lunar surface, some fifty-plus years after astronauts Gene Cernan and Harrison Schmitt closed the hatch on their lunar module and started the journey back home. This is a worthy goal, to be sure, and addresses one of the most salient criticisms of the otherwise brilliant Apollo program: Why, given the expense and effort expended by Americans in those years, were no minorities or women included in the crews? That's an important question domestically, but it's one that resonates internationally as well. And because NASA is now committed to racial and gender inclusiveness, it will be better able to claim a leadership role in space with the developing nations of Africa, Asia, and South America.

Given that the program is estimated to cost some $93 billion over the next few years, though, it's important to understand that Artemis has goals other than equity and inclusion. Artemis is also a response to China's plans to become the world's preeminent space power and the possibility that it will claim large and desirable portions of the moon as its own when it lands its taikonauts there, in the same way that China has claimed large parts of the South China Sea in recent years. The United States wants to forestall being shut out of desirable real estate at one or the other of the moon's poles, which is where water ice is believed to be accessible. But the program is also an effort to rally political support for the notion of free and open access to space resources through the so-called Artemis Accords. This document is a masterfully vague statement of shared political principles and an intent to cooperate authored by the United States and now signed by thirty-nine nations, including Great Britain, Japan, and up-and-coming space power India. Notably, it has not been signed by China, which has partnered with Russia to create a wholly separate coalition of would-be moon miners to support a planned International Lunar Research Station. Among the participants in *this* venture are Pakistan, South Africa, and Venezuela.

And finally, stung by statements over the years that the nation's Apollo moon landings were more stunt than science, a technological dead end, NASA has repeatedly insisted that Artemis will initiate a *long-term* habitation of the moon. The agency has therefore laid out an ambitious scheme in which a space station in lunar orbit will function as a gateway for traffic from Earth to the moon and vice versa. Astronauts will live on the moon, attempt to make water from lunar ice, and possibly begin mining and agricultural operations. If all goes well, the gateway (and this is what it's actually called, by the way—

the Lunar Gateway) will eventually become a staging point for trips to Mars and possibly elsewhere in the solar system.

Even aside from the prospects of lunar mining and water collection, space historian Andrew Chaikin reckons there are three excellent reasons for a permanent station on the moon. First, study of the moon would be like examining a sort of cosmic Rosetta Stone. It would allow us to penetrate deeper into the mysteries of our planet's—and indeed, our solar system's—formation. Second, the moon will be like an Outward Bound school for solar system settlement, a place where we figure out how to live off-Earth while remaining relatively close to home. The moon is a distant place, of course, but the journey is short and simple compared to a trip to Mars. This makes the moon a good place to practice. And third, the view of our home world from the lunar surface would help reinforce humanity's sense that Earth is a fine but fragile place and that we are responsible for and blessed by its good health.

Excellent reasons, all. Now we just need to get there.

America's Eleven Biggest Space Losses

1–3. Accidents involving the loss of human life belong on a different scale altogether than mishaps that lead to the destruction of hardware alone. America's biggest heartbreaks so far have come with two shuttle disasters and an Apollo testing accident. The *Apollo 1* calamity in January of 1967 was NASA's original sin. Three astronauts died on the launchpad as a result of a fire that flashed through the pure-oxygen atmosphere of their sealed capsule. The risks presented by the welter of electrical wiring in a high-oxygen atmosphere were clear and could have been eliminated—but weren't. "It was like we murdered them, almost," said a mournful Chris Kraft. In the *Challenger* tragedy, faulty O-rings and a terrible decision led to the deaths of seven astronauts just after launch on a frigid Florida morning in January of 1986. Among them: America's Teacher, Christa McAuliffe. The loss of space shuttle *Columbia* during STS-107 in February of 2003 is the gut punch no one talks about. Falling foam insulation damaged the orbiter's left wing at launch. The damage proved catastrophic upon reentry, when superheated gases compromised the wing structure and sent the ship into a fatal spin. Suddenly we had seven more souls to mourn. And the space shuttle's days were numbered.

4. Defunding of NERVA, 1973. Discontinuing research and testing of nuclear-powered rocket propulsion was a short-sighted cost-saving measure under the Nixon administration. Many observers now think that nuclear propulsion of some sort will be needed to push human exploration out beyond the moon, so NASA is starting all over again. Meanwhile, we wait.

5. Platypus-ing of the space shuttle, 1970s. NASA's attempt to create a shuttle for everyone—especially the U.S. Air Force—made the

orbiter bigger, heavier, and more complicated than it needed to be. This is in turn led to higher operating expenses and chronic launch delays.

6. Decommissioning the space shuttle before a substitute could be developed, 2011. Paying the Russians for transportation to space was not a good look for the country that brought the world Apollo, *Voyager*, and Hubble—and it was extremely expensive, to boot.

7. Vanguard TV-3 launch (or lack of launch), 1957. America's first response to Sputnik was a Marx Brothers skit, complete with an exploding rocket and a runaway satellite. The Soviets sarcastically offered to send financial aid to help the United States get off the ground. The offer was declined. And we learned. Twenty years later, the American *Voyager* probes set out on their missions, which have now taken them billions of miles from Earth.

8. *Mars Polar Lander/Deep Space 2*, 1999. America's first attempt to dig for water on the Red Planet ended in a costly failure when the *Mars Polar Lander*, um, *landed*. While the cause of the loss is still not entirely settled, it probably occurred when the probe's legs were deployed as it approached the Martian surface. Onboard computers mistook the vibration from this action as the impact of landing and shut off the thrusters that were meant to slow the probe's rate of descent. The result was a terminal crash. No Martians are believed to have been involved.

9. *Mars Climate Orbiter*, 1999. A forehead-slapping mismatch of metric and U.S. customary measurement units (i.e., "English" units) by NASA and one of its contractors led to computer confusion and the loss of a $125 million Mars probe just as it reached its target.

10. *Ranger 6*, 1964. The Ranger program was created to send camera-bearing probes to the moon to scout destinations for Apollo lunar missions. The first five missions failed and led to a major review and retooling of the project. Thus, hopes were high for the new and improved *Ranger 6*, which had a near-perfect launch and trajectory but then took what one newspaper called a "death plunge" into the moon without sending back a single picture. It was yet another embarrassment

for NASA. Fortunately, subsequent Ranger missions were more successful and somewhat redeemed the program.

11. The "Four-Inch Flight," 1960. Inevitable early launchpad problems like the Mercury-Redstone 1 failure in November 1960 were still embarrassing, partly because they were so keenly watched. This debacle, in which a Redstone launch vehicle traveled less than a foot before settling back on its launchpad, was another body blow to American engineering morale. It didn't do much for the moods of the astronauts who were waiting to ride the rocket, either.

16

The New Space Race

Well funded and capable of the sort of long-term planning that is often
unavailable to NASA, the dynamic and secretive Chinese space program
has emerged as America's chief cosmic rival.

Sputnik sparked the world's first space race, a thirty-four-year, on-again/off-
again cosmic contest between the United States and the Soviet Union intended
to demonstrate the superiority of one sociopolitical system over the other.

Because they could never get their equivalent of the Saturn V rocket, the
N-1, to work, the Russians lacked the firepower to send human beings to the
moon. Thus, in July of 1969, the world watched as *Apollo 11* lifted off from
Kennedy Space Center. A remote-controlled Soviet probe, *Luna 15*, which in
one artist's depiction looks like a potbelly stove resting on a beanbag chair,
took flight around the same time and headed for the same destination. Indeed,
there were concerns that the two vessels might collide. They didn't. Armstrong
and Aldrin did their waltz on the lunar surface. *Luna 15* crashed into it, and
the biggest prize in the space race—a flag on the moon, a bin full of rocks—
went to the United States.

Competition continued nevertheless. Billions of dollars, untold hours,
and huge amounts of intellectual energy went into the contest, which often
seemed like a pointless pageant of technological one-upmanship: you send a
man to space, we send *two* men to space; you do a space walk, we do a *longer* space walk, and so on. There were military dimensions as well. The Soviets armed one of their Almaz space stations with a rapid-fire "space cannon"
and are reported to have test-fired the weapon in orbit at least once. The U.S.
Air Force developed its astronaut maneuvering unit, a hydrogen peroxide-
powered jetpack, with an eye toward sending astronauts outside their capsules to inspect and possibly disable enemy satellites.

The Russians were pioneers in space station deployment and occupation,
and they spent a weird amount of time and effort sending probes to Venus.
But in the late 1980s, when they were openly copying U.S. technology such

as the space shuttle and NASA's nifty manned maneuvering unit, the Soviet space effort began to falter, crippled by a lack of funds. The end came quickly. A cosmonaut who went into space as a Soviet citizen found himself a man without a country when the Soviet Union formally broke up in 1991. Sergei Krikalev was stuck on the *Mir* space station for months and returned to Earth in March of 1992 as a citizen of something called the Russian Federation. The Federation became an ally of the United States in space exploration rather than an antagonist. Indeed, by the dawn of the twenty-first century, the United States and Russia were partners in operating the International Space Station, and American astronauts were traveling to the station and back on Russian spaceships. While a significant amount of good will between the two space programs evaporated when Russia invaded Ukraine in early 2022, Moscow and Washington still cooperate on the ISS.

The importance of the space race has been exaggerated. It was a significant battlefield, to be sure, but there were other competitions as well. In the culture wars, Elvis, blue jeans, and Hollywood westerns took on Ukrainian folk dancers and the Bolshoi Ballet. Jonas Salk developed the polio vaccine in 1953 and elected not to claim any money from the sale of the patent. The United States led the way in vaccinating the world, winning hearts and minds in the process. Americans triumphed economically by selling sewing machines and soft drinks, Chevrolets and jet engines to people and nations in every corner of the globe, improving lives one toaster at a time and proving decisively that capitalism is capable of generating a greater number of gadgets, gizmos, and cardiovascular problems for the world's population than any economic system heretofore devised. In the long battle for hearts and minds, it's unclear whether culture or the cosmos was the more significant battlefield. Neil Armstrong and Cape Canaveral were important advertisements for the United States, but so were Muhammad Ali and Coca-Cola. The fall of the Berlin Wall almost certainly had more to do with the Beatles than with Frank Borman.

The East Is Red

Meanwhile, China was starving.

The nation was born as the People's Republic of China in 1949 after a bloody civil war in which Mao Zedong's communists defeated Chiang Kai-Shek's U.S.-backed nationalist forces. The focus of the new nation's early space effort was the so-called Two Bombs, One Satellite program, with the stated goal

THE NEW SPACE RACE | 183

of developing an atomic (and, later, a hydrogen) weapon, an intercontinental ballistic missile, and a satellite. The Soviets helped for a while, until the two nations fell out over geopolitical matters. Former Caltech Suicide Squad rocketeer Tsien Hsue-Shen, embittered by his treatment by American law enforcement officials during the Red Scare years, provided important leadership. Still, social and political upheavals crippled the nation's scientific community during the fifties and sixties and left China's technological efforts lagging behind both the Soviet Union and the United States. Mao complained at one point that his country couldn't even launch a *potato* into space. But he and his disciples were largely to blame. China is the only nation to have had one of its best rocket scientists beaten to death by ideologically inspired thugs, as Yao Tongbin was in 1968 during the Cultural Revolution, a period of prolonged socialist hysteria in which scientists and intellectuals were brutalized for their real or imagined adherence to "decadent" Western ideas. The incident is echoed in the early pages of Chinese writer Liu Cixin's novel *The Three-Body Problem*. "To develop a revolutionary science," one fanatic shouts as she harasses a physicist for his allegedly reactionary ideas, "we must overthrow the black banner of capitalism represented by the theory of relativity!"

China finally launched its first satellite in 1970, called Dong Fang Hong ("the East is Red"), but its space program languished for lack of funds for some years afterward. It wasn't until the late seventies that the nation was able to work seriously on its space exploration efforts. China sent its first uncrewed Shenzhou spacecraft into orbit in 1999. In 2003, forty-two years after the first American and Soviet spacefarers took flight, a Chinese taikonaut named Yang Liwei entered the cosmos, launched on a Long March 2F rocket and enclosed in a Shenzhou capsule. *Shenzhou* means "divine vessel," harkening back to pre-communist Chinese religious traditions. Long March, on the other hand, refers to Mao Zedong's heroics in leading China's socialist armies in their extended evasive maneuvers against nationalist forces in the 1930s. This mating of spiritualism and socialism, which would have been unthinkable until reformist leader Deng Xiaoping's modernization campaign in the 1970s, is an intriguing sidelight of modern China.

Though the Chinese were late to send citizens across the Kármán line, they have made up for lost time. Things haven't always gone as planned. While this era is now largely forgotten, China and the United States collaborated on a number of military and space projects in the seventies and eighties. Washington was capitalizing on the fact that China and the Soviet Union were at odds following a violent border dispute between the two nations in 1969. This incident ushered in the Sino-Soviet split, a period during which China and the Soviet Union had generally chilly relations and Washington and Beijing meanwhile found a common cause in their shared fear of Soviet aggression. In early 1979 Deng Xiaoping visited DC and was reportedly granted a midnight tour of the nation's top secret nerve center, the headquarters of the Central Intelligence Agency, where he viewed spy satellite imagery of Soviet military installations. According to reporting by the *New York Times*, a joint Chinese-American surveillance operation called Project Chestnut was afterward established in western China to listen for evidence of Soviet missile launches. Later, American aerospace contractors allegedly worked with the Chinese government to improve Chinese space technology, and the Chinese space program offered commercial rocket flights to American interests for a period in the late 1980s and into the midnineties. These vehicles had an uneven performance record. In 1996 for example, a Long March 3 rocket carrying an American satellite veered off course shortly after launch. It exploded in midair, and pieces of it slammed into a nearby village, killing somewhere between five and five hundred residents (the number remains in dispute) and doing substantial property damage. Errors continue to plague the program. Parts of the *Tiangong-1* space station rained down on Earth in April 2018, just as portions of *Skylab* did in 1979, and the Chinese Space Agency has yet to demonstrate that it can control reentry of its big Long March 5 rockets, sections of which have also crashed in alarming fashion over the years.

In just the past few years, though, China has landed a rover on the far side of the moon and another on Mars, transmitted vivid images from the surface of the Red Planet, deployed and assembled its own space stations, and announced plans to send human beings to Mars in the 2030s—all activities that would have been seen as the exclusive domain of the United States at the dawn of the new century. And while the current Chinese space station is smaller than the ISS, it's newer, and it will be around long after its Western rival has plunged to Earth, with no immediate successor in sight. For some, the message is clear: "CHINA IS SERIOUS ABOUT WINNING THE NEW SPACE RACE" opined a columnist for the *Washington Post* in July of 2023. The UK version of the online magazine *Wired* headlined an April 2023 story "CHINA'S BID TO WIN THE NEW SPACE RACE." In November 2023 *Polit-*

ico mentioned warnings about Chinese advances—including work on a suspected space-based nuclear weapon—sounded by both the Pentagon and a congressional advisory body in its story "THE NEW SPACE RACE WITH CHINA." There hasn't been a single Sputnik moment to mark the beginning of the contest, but it's obvious that China aims to become America's equal in aerospace technology, if it isn't already.

Managing the New Competition

Just as in the first cosmic contest, low-Earth orbit is only one arena in a bigger competition. In the years since the disintegration of the Soviet Union, the United States and China have emerged as the two biggest bullies on the geopolitical block. Beijing has a regular army of something like two million soldiers, and its military budget is second only to Washington's. Its hypersonic missile technology is reportedly superior to that of the United States, and a leak of sensitive American intelligence documents in April of 2023 revealed that the Chinese are engaging in what the *Washington Post* called "the largest nuclear buildup since the Cold War." Air force secretary Frank Kendall gave a speech in September of 2022 in which he acknowledged that space has become at least partly militarized. In evaluating American responses to possible space-related threats, he said, "I have three priorities: China, China, China." And the *New York Times* reported in May of 2024 that the Pentagon is spending massive amounts of money to counter what it sees as increasing threats to American satellites posed by Chinese and Russian space weapons. "If we don't have space," said one high-ranking official, "we lose."

There are currently no proxy wars raging between the United States and China, no Vietnams or Angolas, as there were between the United States and the Soviet Union in the Cold War days. But there are looming disputes just the same. Taiwan is a devoted American ally, but its independence continues to rankle Beijing, which considers its offshore neighbor not as a prosperous fellow nation but as a wayward province. China claims sovereignty over a huge portion of the South China Sea, despite the protests of a number of other nations, including Vietnam, Malaysia, the Philippines, and the United States. Accusations of spying and industrial espionage have troubled relations between the United States and China for years. China is notoriously indifferent to patent and copyright laws, and suspicions that the Chinese were pirating U.S. technology for their space and defense programs led to the passage of the Wolf Amendment in 2011, which essentially prohibits NASA from working with communist Chinese interests on any and all aerospace projects. This is a prime reason why no Chinese spacefarer has visited the ISS.

The two nations are ideological as well as economic rivals. Though today's China is not the Soviet Union of 1957, with its threats to "bury" capitalism and its practitioners, there is still tension between the two systems. The choice is less between two competing ideologies than it is between two practical examples. China is the world's leading manufacturer. It's a sprawling but mostly stable country that viciously quashes dissent and meaningful political discourse but seems to be able to Get Big Things Done. The United States is the world's second largest manufacturer. We continue to lead the world in scientific and technical innovation. Our economy generates huge amounts of wealth for some but works less well for others, and the nation seems increasingly venomous in its political discourse and distressingly inconsistent in its foreign relations.

The Sino-American competition is especially keen in Africa, where the Chinese have spent billions of dollars in funding and constructing infrastructure and cultivating influence with local governments pursuant to its ambitious Belt and Road Initiative. The United States has countered with the so-called Blue Dot Network of infrastructure investment initiatives, named after *Voyager 1*'s *Pale Blue Dot* photograph of Earth and engineered primarily by the United States, Great Britain, Japan, and Australia. The network assesses proposed infrastructure projects in the developing world and issues qualitative ratings of the projects based on factors such as how well they promote market-driven economic growth and demonstrate resilience to projected climate change and other environmental challenges. The aim is to encourage and channel private investment in smart, sustainable development. Just as during the Cold War, off-planet excursions can be viewed as a small part of a much larger struggle for influence around the globe.

Space will not be explored by the poor or powerless. Russia, Canada, and the European Space Agency have the brainpower to mount cutting-edge space missions, but they lack funds. India has both, but it is just starting to flex its rocket-building muscle, while Japan has so far chosen modest objectives for its own technologically advanced missions. A handful of private companies have the technical ingenuity and maybe soon the hardware to mount ambitious space operations, but it remains to be seen how willing their owners or investors will be to finance a long-term exploratory or settlement mission with no immediate promise of profit. (SpaceX and Blue Origin may be exceptions here, as their owners have exceptionally deep pockets, and the companies were founded with the express intention of facilitating human habitation either on Mars or within large space stations.) The U.S. military has significant space assets as well; indeed, the U.S. Space Force has a solar-powered spacecraft, the x-37b, that is capable of staying in orbit for months at a time. But the space force's job is to protect American assets. It has neither the budget nor the authority to undertake exploratory missions.

China and the United States are currently the only two nations that have the money and know-how to mount significant exploration efforts. China has the additional advantage of longer timelines and steadier direction for its space initiatives. Unlike in the United States, where politics tends to sway space policy from administration to administration, the Chinese have the luxury of stability. The United States, on the other hand, can draw on the creativity and energy of a much less controlled technological culture. It will be fascinating to see how the interplay and advancement of the two space programs play out. There is bound to be military involvement, mutual spasms of distrust, and occasional saber rattling. The ability to disable a country's global positioning and communications satellites is the power to blind that nation and its military forces, something neither the United States nor China will allow the other to do without a fight. Thus, there is a distinct risk that the new space race could reflect or even exacerbate political and military tensions between the two countries. Human beings tend to think in binary terms. Us vs. Them. Good vs. Evil. No doubt the prospect of a Chinese land grab (*moon grab?*) at the lunar south pole or the capture of a mineral-rich asteroid for the sole use of the Chinese Communist Party will be used as motivations for both NASA and DOD space projects in the future.

But we don't have to fall into the tit-for-tat thinking of the Cold War. Not every action demands a counteraction. And perhaps we can further insulate ourselves from another binary competition by encouraging the space efforts of other nations as well, so that a moon grab by China is an offense against multiple countries with interests in lunar resources, not just ours. This is the context in which assistance to and cooperation with, for example, the Indian, European, and Japanese space programs makes good sense. Finally, we should remember that a rival today can be an ally tomorrow. The Soviet Union was a bitter enemy for years after World War II, a friend for thirty years after that, and is now something of both. We were friendly with the Chinese for years, until we weren't. Things change. Politics is no exception.

Risk can be managed. Competition isn't necessarily conflict. Nor is it necessarily negative. After all, it was competition with the Soviet Union that spurred the development of important technologies during the first sixty years of America's space program. No lives were lost in space due to military adventurism on either side. And whereas the first space race was a two-horse affair, the current scramble to get a foothold on the moon has several participants—including even some private companies. The goal of all nations in space should be global good rather than partisan territorial or ideological gain. For example, what if the United States and China competed to see which country could best protect the earth from rogue asteroids? Both nations have announced plans for doing exactly that; NASA has actually demonstrated the viability of its proposed technique. What about safeguarding the environment through satellite detection of chemical pollutants and polluters? Which space station, China's or the ISS, will prove to be the bigger lure for astronauts and scientists whose own nations can't afford to maintain a presence in low-Earth orbit on their own? And which country has the best ideas and machinery for cleaning up the planet's granular hula hoop of space junk?

Winning the hearts and minds of the world's population through technology, exploration, and inspiration is a perfectly legitimate goal of a new space race. Winning territory or political concessions through threatened or actual force, whether on earth or in the cosmos, is not. With these parameters in mind, let the race begin—or, maybe more appropriately, given recent Chinese accomplishments, let it continue.

17

The Commercialization of Space

At long last, private industry is building increasingly capable and ingenious
space vehicles. But not every shiny new space company will survive.

NASA partnered with private interests early on. Its first commercial space
launch took place in July 1962, when, for $3 million, the agency launched
AT&T's pioneering telecommunications satellite, Telstar 1, on a Thor-Delta
rocket. The world had seen satellites that "echoed" ground transmissions
before. In fact, the U.S. Navy used the moon as a passive radio-wave relay station in its Operation Moon Bounce in the late fifties and early sixties. NASA
launched and operated a giant mylar balloon satellite called Echo 1 in 1960
that redirected microwave communication signals, and a Pentagon project
called West Ford attempted to scatter hundreds of millions of hair-like copper filaments in orbit in 1961—and then again, more successfully, in 1963—
to provide greater radio wave reflectivity. But Telstar was a new thing, a tiny
spacecraft that not only reflected terrestrial signals but also amplified them
before sending them along, using its altitude to avoid the problems caused in
trying to relay radio waves along the curved surface of the earth.

Telstar 1 was a sphere, eighty-eight centimeters in diameter and weighing
in at 170 pounds. Most of its surface consisted of solar cells, which turned
sunlight into a total of around fifteen watts of power to run the satellite's
transponder, radio receiver, and transmitter. The cells were covered in a thin
layer of artificial sapphire to protect them from cosmic radiation. Telstar,
which looks remarkably like a miniature version of the Death Star of *Star
Wars* fame, or possibly the world's first disco ball, relayed the first live television transmission—video of the American flag flapping in the breeze—from
the United States to Europe on 11 July 1962.

Nor was the spacecraft itself the only futuristic wonder. Telstar 1's prin-
cipal ground station, in tiny Andover, Maine, became an overnight tourist
destination, visited by up to two thousand sightseers a day interested in Telstar's huge "mechanical ear," which stood seven stories high and weighed 340

tons. The antenna was hangered inside the world's largest inflatable structure, a futuristic spherical radome that was held up by air pressure. Unsure what a radome is? Think of those golf balls of the gods you see sometimes near airports and military installations. They're large, usually white, sometimes multifaceted, and served to shield sensitive communications equipment from the elements but not to block radio signals while doing so. How sensitive was the giant antenna, one tourist asked? A helpful employee responded that it could hear a flea flap its wings in far-off Bombay. (How this metric was calculated was not disclosed—nor was the fact that fleas don't actually have wings.)

Telstar was a "scientific celebrity" in its day, not quite as important as Sputnik perhaps but still an impressive demonstration of American technological prowess. The legendary and ill-fated British music producer Joe Meek even wrote a song about it. The Tornados released Meek's instrumental composition "Telstar" in August of 1962. Featuring an analog electronic synthesizer that sounded a little like a theremin, "Telstar" was freaky and futuristic but somehow jaunty as well, like a pep rally on an alien but generally congenial planet. The record sold some five million copies and became a No. 1 hit in both the United Kingdom and the United States—the first U.S. No. 1, in fact, by a British group. Unfortunately, the Beatles began scuttling up the pop charts soon afterward, and young music fans had to give up their reveries regarding orbital communication satellites in favor of more mundane concerns regarding whether particular human beings, often teenagers, loved each other, for how long that love would last—if indeed it existed in the first place—and how intensely such love could be perceived in the absence of a 340-ton antenna to help with the measurements.

Telstar 1 is still orbiting Earth, though it has long since ceased to function. Ironically, the satellite's lifespan was shortened almost immediately, thanks to an American high-altitude nuclear detonation that took place shortly before the satellite's launch. The test, called Starfish Prime, produced an explosion so powerful that it created an artificial radiation belt that damaged not only Telstar but also a number of other satellites. No matter. The first Telstar was succeeded by a second, and with additional descendants—all of which have been sent into geosynchronous orbits, high-altitude flight paths that allow an object to stay in more or less the same position in the sky as the earth rotates. The current iteration of Telstar is Telstar 19, launched on a SpaceX Falcon 9 rocket in 2018.

Once a novelty, telecommunications satellites are now used every second of every day for radio, television, internet, telephone, and military applications. They're launched on commercial rockets on a regular basis and have become commonplace—indeed, almost a nuisance. SpaceX, for example, has already deployed thousands of its little Starlink broadband satellites and shows no signs of slowing down. The Starlink network has swollen the ranks of the hundreds of other manmade chatterboxes roaming the skies. Thanks to the work of private enterprise in space, we've never been more connected.

From Big Ideas to the New Space Economy

Despite the dearth of crewed spaceflights in the late seventies, interest in human habitation of the heavens swelled. This was in large part due to the work of Gerard K. O'Neill. In fact, it's fair to say that it was O'Neill, together with astronomer Carl Sagan, who kept American interest in space exploration alive in the lackluster years after Apollo.

Sagan was a television personality and portable pundit, a cosmic Pied Piper of immense imagination and wit who delighted in seeing echoes of earthly experience in the farthest reaches of the universe. He was, in short, a poet. He grew up in the Bensonhurst neighborhood of New York City in a family with few resources and was first intrigued by the natural world in what he found in books and museums. During his studies at the University of Chicago, he formed a lasting interest in planetary science. He earned an assistant professorship in astronomy at Harvard University but was eventually denied tenure—a questionable decision, to say the least—and thereafter moved to Cornell, where he taught for the rest of his life. He was a prolific writer and speaker who seemed to be equally at ease with both the astronomer Gerald Kuiper and the comedian Johnny Carson, and he never stopped advocating for the importance of wonder and awe in the face of the universe's mysteries. He was an early proponent of SETI, the search for extraterrestrial intelligence. He hosted an immensely influential space-based television series called *Cosmos* on PBS, and he wrote both a Pulitzer Prize–winning nonfiction book about human evolution and a best-selling novel about the day Earth is contacted by aliens. A slight man with a winning grin and boundless enthusiasm, he was affectionately mocked in the media for his favorite phrase: "billions upon billions," simplified by comedians to "billions *and* billions," delivered with a rubbery, bombastic *b*. By 1980, Sagan seemed to be everywhere, a pitch man for infinite possibilities.

THE COMMERCIALIZATION OF SPACE

Gerard K. O'Neill was nowhere near as familiar a public figure as Sagan, but he was tremendously influential nonetheless. He wasn't content with echoes of Earth and glimmers of immortality. In essence an engineer, practical and plain spoken, he wanted to seal pieces of the planet in translucent packages for off-world consumption. Boyish, with thin lips and an unfortunate bowl cut that gave him a Mr. Spock–like look, O'Neill was not only a big dreamer but also a brilliant scientist. He earned his PhD in physics from Cornell in 1954 and went on to teach at Princeton, where he did pioneering work in particle physics. Eventually, though, he looked upward. He focused on the idea of creating self-supporting cosmic habitats at the L4 and L5 Lagrange points—locales in space aligned with the moon and Earth in such a way that objects can be placed in stable "halo" orbits there.

In his 1976 book *The High Frontier: Human Colonies in Space*, O'Neill wrote about his ideas for mining ores on the moon and the ways in which massive, self-sustaining, solar-powered sky cities could be constructed, maintained, and populated. Earth was breaking, he argued. Humankind was killing it through overpopulation, pollution, and rapid consumption of its natural resources. To save the planet and ourselves, we needed to take to the heavens. While he was initially hopeful that NASA would sign on to develop his habitation ideas, O'Neill and his supporters in the new L5 Society gradually grew discouraged about the prospects of the government undertaking such mega-projects. But this was okay, they concluded. Maybe government wasn't the best way to get important things done anyway.

The L5ers eventually merged with another advocacy group, the National Space Institute, to form the National Space Society. While popular interest in O'Neill's ideas gradually waned, members of the old L5 Society went on to influence American space policy—and still do today. They continue to preach the gospel that private interests can create space technology cheaper and faster than the bureaucracies of the federal government and the legacy aerospace contractors that thrived during the big-budget, whatever-it-takes days of Apollo. Perhaps the most prominent O'Neill acolyte active in the space industry is Jeff Bezos. Another influential former L5er—a "space pirate," as she calls herself—is Lori Garver. As deputy chief of NASA under President Barack Obama, Garver did as much as anyone to push NASA in the direction of partnership with new and innovative space companies like Mars-boosting bad boy Elon Musk's SpaceX. The commercial space market—including crew and payload delivery missions, satellite manufacturing and maintenance, and imaging, geo-location, and broadband services—is now worth trillions of dollars. It will only get bigger.

Conestoga 1

One of O'Neill's most ardent acolytes was a man named David Hannah, a mild-mannered, middle-aged Houston real estate investor who for most of his life displayed no particular interest in science or space exploration. This changed suddenly. In July 1976 Hannah read about O'Neill's work in a copy of *Smithsonian Magazine* and was overcome by the futurist's logic. He came to feel that God and O'Neill had ordained him to build rockets, both to further the O'Neillian vision and to demonstrate the off-world potential of free enterprise. The real estate man eventually formed a company called Space Services, Inc. of America (SSIA), attracted investors from among Houston's old-line big money men, and hired a rocket designer.

SSIA's first rocket, the Percheron, was an elegant liquid-fueled creature that blew up on the launchpad. In his second attempt, Hannah went with a less sexy but more reliable solid fuel missile—a repurposed Minuteman, purchased from NASA. He brought in a crew of experienced American and German rocketeers and hired steely-eyed Deke Slayton to ride herd on the whole operation. Bringing the grizzled ex-astronaut on board was a smart move. Slayton was fully capable of scaring the rocket into flight, if need be. Preparations came with the usual assortment of headaches and glitches. But on 9 September 1982, Hannah, Slayton, and SSIA launched their Conestoga 1 rocket from a cow pasture on Matagorda Island, on the Texas Gulf Coast. As Stephen Harrigan of *Texas Monthly* reported:

> The firing of the *Conestoga I* was referred to on the countdown sheet as the ignition event. This event began, of course, at T minus zero, when an explosive squib at the top of the booster was detonated, sending a shower of tiny burning pellets into the hollow, star-shaped core of the rocket motor. The pellets created a gas, and the gas reacted so swiftly and so furiously with the solid fuel surrounding it that the *Conestoga I* began to move upward into the air.
>
> The sun struck the flame coming out from the bottom of the rocket with such intensity that the flame had a metallic gleam and seemed to be all of a piece with the vehicle it was thrusting upward.

"We have ignition," [flight controller] Sallie Chafer said, a little calmer now. "The *Conestoga I*, the world's first free enterprise rocket, is on its way."

The rocket was well above the horizon before the rumble of its passage was audible. It left a white contrail that seemed to veer and twist crazily on the ceiling of the sky. The rocket itself was visible for about thirty seconds, then it melted imperceptibly into the atmosphere, and that was the last anybody ever saw of it.

The bird climbed flawlessly to an altitude of 192 miles. It released a dummy payload—forty gallons of water—and then began to fall back to Earth. Conestoga 1 landed somewhere in the Gulf of Mexico, some 320 miles downrange of Matagorda. SSIA thus became the first private company to put a rocket into space. It was a major story at the time, portrayed as a blow for private industry and entrepreneurship struck by a bunch of gutsy no-names (Slayton excepted, of course) trying to do what only America's mammoth space agency was thought able to accomplish. Newspapers around the country ran stories with headlines like "LAUNCH SCORES ONE FOR FREE ENTERPRISE." One observer called Matagorda Island the "Kitty Hawk of the commercial space industry."

Though the Conestoga 1 story is little known today, there's some truth to the phrase, and the SSIA story would be repeated over and over again in the coming years as brash rocket companies stepped up to the plate to challenge the commercial launch market dominated by NASA and the European Space Agency. Other aspects of the story would be repeated as well. Hannah had an ambitious agenda for the launch of satellites and scientific instruments but ran into operational and budgetary problems. The company's last gasp came in 1995, when it attempted to send a NASA payload into space from Wallops Island, Virginia, using a Conestoga rocket festooned with additional boosters. Slayton was no longer at the helm, as he'd died of brain cancer two years earlier. He wouldn't have been pleased by the result. The launch of SSIA's new Conestoga 1620 ended in disaster when the rocket broke apart only forty-six seconds into flight. It was David Hannah's final launch. SSIA folded not long afterward.

The Dawn of the Space Economy

Also in 1982, longtime NASA designer Max Faget and others started a company called Space Industries, Inc., with the idea of launching a private space station called the Industrial Space Facility. The plan was that once deployed, the facility could be leased to private industry to perform experiments or manufacture products in microgravity. The start-up gained favorable attention from President Ronald Reagan. He wanted to fund it, but Congress demurred, and the company's central goal was never achieved. Space Industries, Inc., was thus a charter member of a still-expanding society of failed space start-ups, which has come to include such notable efforts as SSIA, the American Rocket Company, Kistler Aerospace, Rotary Rocket, Beal Aerospace, and Bigelow Aerospace.

The White House began encouraging the creation of commercial space vehicles in the 1980s, partly in response to the examples of SSIA and Space Industries, Inc. Despite Reagan's folksy, Roy Rogers–like appeal, he was a bit of a space freak. His wife Nancy looked to the heavens as well. She was a devotee of astrology and in fact had a personal astrologer to help her understand the occult mysteries of the stars. Reagan focused on technology. Even as his budget cutters loped around Capitol Hill, snipping and gouging, Reagan proposed hugely ambitious (and expensive) space plans, like the laser-studded Star Wars Defense Initiative and the ambitious space station *Freedom*. NASA created the Office of Commercial Programs in 1984, as the space shuttle program seemed to be hitting its stride. And in October of that year, Reagan signed into law the Commercial Space Launch Act, designed to encourage the development of an American industry of private operators of expendable launch systems.

Such developments put NASA in an awkward situation. While no one at the agency wanted to buck the president's enthusiasm for free enterprise and market competition, the fact remained that NASA needed as many customers as it could get in order to meet the agency's highly optimistic financial projections for the space shuttle. Thus, while space administrators nodded along with the free-market talk and loved the idea of entrepreneurs bringing new satellite and space science payloads to the shuttle, NASA was notably mum on proposals to help private investors get into the launch vehicle game. Economic forces began to undercut the shuttle's monopoly position anyway. The European Space Agency (ESA) started to compete with NASA for commercial satellite launches in 1984. Using its comparatively simple Ariane rocket, ESA soon developed a healthy customer base among companies, even U.S. companies, that felt that NASA's services were too expensive and too often subject to delay. American entrepreneurs saw a potential market for themselves in lower-cost commercial launch services and began knocking at the doors of Congress to expand the abilities of private interests to either build and operate their own rockets, or to buy surplus military technology to accomplish such launches.

The first surge in commercial activity in space came in the wake of the *Challenger* disaster in 1986. The space shuttle was grounded for years afterward. When the program started up again, NASA shunted off routine satellite deployment activities, instead prioritizing missions with payloads specifically designed for shuttle transportation or requiring not boots on the ground but hands in the sky (the Hubble Space Telescope launch and repair missions, for example). Private companies began to compete in earnest for commercial space launches, with United Launch Alliance, a joint venture of long-time aerospace giants Boeing and Lockheed Martin, taking an early and government-encouraged lead.

Nevertheless, and despite the fact that everyone predicted it was coming, it took a long time for a competitive commercial rocket industry to develop. Space tourism initially seemed like the most obvious route to profitability. Airline giant Pan Am started offering reservations for the chance to buy tickets to the moon in 1968. Its rival, TWA, soon followed suit. Space shuttle contractor North American Rockwell studied the feasibility of adding a seventy-four-person passenger compartment to the orbiter's payload bay in the eighties, and in the decades afterward a number of newborn rocket companies like Armadillo Aerospace and XCor launched themselves through clouds of newspaper headlines with big promises regarding space tourism—and almost as quickly disappeared. In 1985 a Seattle company solicited customers willing to shell out a million dollars for an inaugural forty-eight-orbit sightseeing tour of space, which it expected to become available sometime in 1995. A major milestone in commercial spaceflight occurred in 2004, when a team led by long-time airplane designer Burt Rutan won the $10 million Ansari X Prize for designing and successfully demonstrating a crewed vehicle that could fly to space and return to Earth twice in the course of two weeks. Rutan and British entertainment and airline mogul Sir Richard Branson thereafter joined forces to create a larger version of Rutan's craft for use by Branson's fledgling space tourism company, Virgin Galactic.

Despite all the hype, high-altitude tourism has yet to become a major industry. Payload launch services have by contrast proven much more lucrative. During the George W. Bush and Obama administrations, NASA finally moved to encourage entrepreneurial space operations in a big way through milestone-based contracts for development of commercial space vehicles, along with the award of substantial, multi-mission contracts for delivery of, first, cargo, and then human beings to the International Space Station. It was a major shift for NASA to abandon its traditional insistence on building its own space vehicles, and it was prompted at least in part by the realization that once the shuttle was retired, as it was in 2011, there would be no U.S.-built spacecraft to handle flights to the ISS. This change in approach, along with the dogged efforts of aerospace investor/owners with extremely deep pockets and radical increases in computing power and availability, opened the door for companies like SpaceX to become meaningfully integrated into America's space program. SpaceX for one has never looked back. In the past decade and a half, it has become the world's largest and most successful commercial launch company.

NASA's stated position now is that it leaves more or less "routine" space activities to commercial interests. Routine doesn't necessarily mean easy. It's an axiom of aerospace engineering that space is hard. But technologies and know-how now exist that make low-Earth orbit accessible to private industry. NASA, meanwhile, has reserved to itself more challenging scientific and exploratory missions. Some examples: construction and deployment of the James Webb Space Telescope, the *New Horizons* probe's visit to Pluto, and the Double Asteroid Redirection Mission, NASA's successful attempt to determine the feasibility of knocking a potential Earth-threatening asteroid off course and into deep space.

None of these missions offer the necessary incentives to private enterprise, which has to at least bow in the direction of profitability at some point. But in areas where profit is possible, commercial aerospace companies have slowly but assiduously expanded their operations in space. They have been encouraged by federal legislation. The Commercial Space Launch Competitiveness Act, sometimes referred to as the Spurring Private Aerospace Competitiveness and Entrepreneurship (SPACE) Act of 2015, is an update of U.S. law regarding commercial space use. The new law explicitly allows U.S. citizens and industries to "engage in the commercial exploration and exploitation of space resources," including water and minerals, to be found on the moon, other planets, and asteroids. The act recognizes that there are riches to be found in space, and it encourages private sector prospectors to figure out how best to harvest them. While there are currently no space mining operations underway, there are, in fact, companies—California's TransAstra Corporation, for one—trying to figure out where to start.

It's difficult to summarize the players in the commercial space field, because there are so many of them and their activities are so varied. Some companies make satellites. Others want to mine asteroids. Still others are trying to figure out how to build space stations. Paragon Space Development Corporation designs spacecraft life-support systems, specializing in techniques for recycling urine and sweat into drinking water—or, as astronaut Don Pettit puts it, turning yesterday's coffee into *today's* coffee. SpaceBorn United is trying to figure out how to create embryos in low-Earth orbit. Celestis launches people's earthly remains—their *ashes*, that is—into the heavens in a sort of afterlife space tourism; Gene Roddenberry and Dr. Gerard K. O'Neill have already made the voyage.

Several space companies are so new that they have no real track record to consider. Take BluShift Aerospace, for example, which wants to make eco-friendly rockets powered by "an organic substance"—provenance unknown—discovered on a Maine farm. Relativity Space, meanwhile, uses 3-D printers to create the components of its rockets. As intriguing as the ideas are, neither company has actually launched anything into space yet. With the importance of *performance* in mind, here is a doubtless soon-to-be-outdated list of a few of the major players in one segment of space commerce, commercial rocket-building operations, and their accomplishments thus far. A couple of these companies will likely survive and grow strong. One appears to be unstoppable. Still others—well, blink and you may miss them.

Astra

A startup building small rockets that launch out of rural Alaska, Astra notched its first successful test flight in 2021, putting a dummy satellite into orbit. Astra is one of a slew of companies that plan to use lightweight rockets to make frequent trips to space to drop off satellites for whomever can spring for the payload. SpaceX's Falcon 9 rockets are used to haul large satellites, batches of satellites, or NASA astronauts into orbit. They stand some two hundred feet tall, or roughly the height of four of Astra's rockets stacked on top of each other. The idea behind companies like Astra is to create smaller rockets that haul less mass into space and can be built cheaply and launched quickly. If Astra succeeds, it could become the FedEx of space delivery. But it's a big *if*. As this book went to press, the company's survival seemed to be a 50/50 proposition, at best.

Blue Origin

Amazon founder Jeff Bezos's company has been less visible than its chief rival, SpaceX. But it has launched several "tourist" flights, one featuring female aerospace pioneer Wally Funk and another carrying *Star Trek* legend William Shatner, on the company's New Shepard rocket. Blue Origin launches and lands its spacecraft in far west Texas, near Van Horn. Its planned orbital New Glenn rocket will be heavier and more powerful than the New Shepard and allow Blue Origin to compete for orbital launches of commercial payloads. Blue Origin has some of the same ambitions SpaceX has, but it has lagged behind its better-known rocketry rival and almost lost out on an important NASA lunar lander contract as a result. That said, the company has demonstrated a nice flair for publicity—*see Shatner and Funk, above*—and has seemed to find its stride of late with its successful tourist jaunts. Like Musk, Bezos has said that he founded his company to get humanity off Earth and onto Mars, although, unlike Musk, Bezos has also discussed building O'Neillian space settlements.

Northrop Grumman

One of the legacy companies that Blue Origin and SpaceX were founded to compete with, Long Island–based Northrop Grumman has an impressive space pedigree that includes production of the Apollo program's ungainly but effective lunar module. The company's acquisition of Orbital ATK in 2017 gave Northrop Grumman a bona fide next-generation spaceship subsidiary and an avenue to participation in resupply of the International Space Station, which it does with the Cygnus spacecraft, a cargo ship launched on either the Northrop Grumman Antares rocket or the SpaceX Falcon 9. The "enhanced Cygnus" vehicle currently in use looks like a chimney on wheels, or a giant camera lens outfitted with two pizza cutters. In 2019 NASA selected a design based on Cygnus as its model for a lunar gateway space station to be built in connection with its Artemis program.

Rocket Lab

Originally a New Zealand company but now based in California, Rocket Lab has established itself as a commercial launcher of small satellites and CubeSats with its Electron rocket. It is perhaps most notable for its June 2022 launch of NASA's CAPSTONE satellite, the American probe meant to determine the feasibility of a halo orbit around the moon for possible use in lunar habitation projects. The company has announced plans to produce spacecraft for human transport and is tabbed as an up-and-comer in the space industry.

Sierra Space

The spin-off of privately held Sierra Nevada Corporation, Sierra Space is the creator of the Dream Chaser vehicle, a shuttle-like spacecraft that has been in development for over a decade. Sierra hoped to obtain a contract for ferrying astronauts back and forth to the ISS. These hopes were dashed when NASA contracted with SpaceX and Boeing instead. The company did receive a contract to provide cargo deliveries to the ISS, though, and at one point stated that the first of such missions would take place in 2022. While this didn't happen, Sierra Space claims that the cargo version of the vehicle is poised to make its debut flight soon. Both the crewed and cargo versions of the Dream Chaser are to be launched by *rocket*—at this stage, ULA's Vulcan rocket, for the cargo version—and return to Earth and land like the shuttle orbiter.

SpaceX

The eight-hundred-pound gorilla of the new space companies, SpaceX was founded in 2002 by the odd, impatient, and occasionally brilliant Elon Musk. Restless and supremely self-confident, often spotted in an OCCUPY MARS T-shirt, Musk attracts the sort of interest and attention in the business world not seen since the days of Howard Hughes. A fervent believer in the necessity of establishing multiplanetary habitations for the protection and preservation of the human species, Musk started his rocket company with the goal of reducing space transportation costs in the short term and enabling the habitation of the Red Planet at some point in the future. Beginning with several unsuccessful launches on a desolate South Pacific Island inhabited mostly by crabs, SpaceX hit its stride with the fourth test launch of its Falcon 1 rocket in 2008, its first to go orbital.

With the help of talented executives like Gwynne Shotwell and engineer

THE COMMERCIALIZATION OF SPACE | 201

Tom Mueller, SpaceX quickly managed to slash the cost of getting payloads into low-Earth orbit and thus carved out a dominant place for itself in the commercial space market, startling seasoned observers and creating legions of both dedicated fans and vocal critics in the process. Musk provided generous infusions of his own capital along the way. This helped. But SpaceX's success wasn't just a matter of money. The company's aggressive innovation, its willingness to blow things up and start over again, and its eerily beautiful technology— *Tail-first booster landings at sea! The biggest rocket ever assembled!*—have combined to make it more than a space vehicle-construction company. It's now a bona fide pop culture phenomenon, notable for the beauty and high-tech choreography of its launches—and its returns.

Retired air force lieutenant colonel Gary Minar worked for NASA for years. He now lives in central California, not far from Vandenberg Air Force Base. He likes to ride his mountain bike as close as he can (around six miles) to the Vandenberg launch site and watch the SpaceX Falcons fly. One recent launch, he wrote,

> was right on time. One million pounds of propellant go through those engines in about five minutes as the rocket ascends. Then, having delivered its payload to space, the rocket turns around. After it burps some fuel to the pumps, it relights three engines and then comes back above the launch site. Then it begins to fall like a bat out of nowhere and again relights the three engines to slow it to multi-sonic speed. When it is around twenty miles high, it lights the landing burn and in the final few seconds goes subsonic. From where the public can view the event, it appears that the sonic BOOM occurs just as the booster touches down. This all happens in seven and a half minutes! The brilliance of the Falcon team is beyond praise.

SpaceX manufactures the Falcon 9 and Falcon Heavy launch vehicles, the Merlin rocket engine, and the Dragon cargo and crew spacecraft. Its achievements include being the first private company to put a liquid-fueled rocket in orbit; to launch, orbit, and recover a spacecraft; and to send a spacecraft to the ISS. Most importantly, in 2020, with its Crew Dragon Demo-2 mission, SpaceX delivered *astronauts* to the ISS. This flight finally ended America's dependence on the Russian space program for transportation to the ISS after the Space Shuttle was retired in 2011. And while Boeing was granted a similar

THE COMMERCIALIZATION OF SPACE

contract for creation of an astronaut-shuttle service, the Seattle, Washington-based legacy contractor stumbled out of the gate. As of this book's publication date, Boeing *still* hasn't managed to get its Starliner space vehicle fully operational, while SpaceX has mounted several successful ISS crew shuttle and return missions.

SpaceX has flown its Falcon 9 series of rockets over three hundred times. The company is developing a broadband satellite "mega-constellation" called Starlink to provide commercial internet service and has so far deployed thousands of the little satellites. SpaceX is also developing Starship, the ambitious and purportedly fully reusable super heavy-lift launch system that it claims will be suitable for interplanetary spaceflight. Referred to by Musk at one point as the Mars Colonial Transporter, the rocket has a stainless-steel hull and an upper stage that looks a little like the head of a flatworm. Starship is intended to become the primary SpaceX orbital vehicle once operational, supplanting the existing Falcon rocket fleet. Indeed, it may well supplant NASA's Space Launch System as the national heavy-lift rocket of choice. According to Musk, Starship will have the highest payload capacity of any orbital rocket ever built. Its first launch took place in April of 2023 and ended in a spectacular tumbling crack-up not long after the spaceship passed "max q," the point of maximum aerodynamic pressure on the craft. The liftoff also damaged the SpaceX launchpad and sprayed debris over a wide area, generally seen to be undesirable side effects of space vehicle departures. Additional test flights have gone better, but also ended prematurely. These were setbacks for the company, but if history is any guide, they were only temporary.

While Starship is a potential rival of the U.S. space agency's Space Launch System rocket as a heavy-launch vehicle for deep-space exploratory missions, much of the relationship between SpaceX and NASA has been cooperative rather than competitive. NASA's investment in the company early on saved SpaceX from bankruptcy, and NASA has been one of Elon Musk's best customers. Indeed, a Falcon 9 rocket lies on one side just outside the entrance to the Johnson Space Center's visitor center. SpaceX is now an important part of the American space program. A privately held company, it is valued at something like $210 billion. It has sent more than twenty cargo and thirteen crewed missions to the ISS, transporting numerous NASA and private astronauts to and from the orbital station. Such is its prominence in the commercial launch space that the *Wall Street Journal* wrote in July of 2023 of SpaceX's "de facto monopoly" on this segment of the industry.

SpaceX is, in short, a force to be reckoned with.

SpinLaunch

This California-based company launches rockets by tossing them into the atmosphere using an innovative centrifugal spinning device. The machine, shaped like a 165-foot-tall garden snail, works like a sling. Standing in New Mexico's Chihuahuan Desert near White Sands Missile Range, it spins its projectile at "many thousands" of miles an hour before releasing it upward, where a booster rocket kicks in to provide additional power. Payloads will likely be quite small—but so, too, says the company, will the launch costs. SpinLaunch is mentioned here not for its commercial activities—it doesn't have any yet—but for its innovative approach to the problems associated with launch.

United Launch Alliance

A joint venture of legacy aerospace contractors—and former bitter commercial rivals—Boeing and Lockheed Martin, United Launch Alliance provides launch services with its Atlas and next-generation Vulcan rockets. The "big boys" of rocketry, ULA benefits greatly from government contracts, awarded partly with the idea that the United States needs (or *needed*, at one point) at least one large, stable, experienced commercial launch provider. ULA was long seen as the gold standard among commercial rocket launchers, safe and reliable but not cheap—and not particularly prompt, either. ULA's workhorse Delta launch vehicle, which had been around in several variants since 1960, was retired in April 2024.

Virgin Galactic

Media mogul Richard Branson's space tourism company was at one time the brightest of the young industry's stars, an enterprise working with proven technology, deep pockets, and a charismatic billionaire owner. But Virgin Galactic was a late bloomer, falling behind the efforts of rival rocket tycoons Musk and Bezos. The company's biggest stumble came when its first spacecraft, called VSS *Enterprise*, broke up in flight and crashed in October 2014. The craft's copilot died, and the pilot was seriously injured, surviving only because of his parachute. After a considerable lag in activity following this crash, the company launched an inaugural low-Earth orbit tourist flight in July 2021, with Sir Richard himself riding along. While Branson thereby won a space tourism battle with Bezos, the flight drew the scrutiny of the Federal Aviation Administration as a result of indications that the spacecraft deviated from its flight path—a sign of either lack of diligence or a mechanical malfunction.

The company got back on track in July 2023, when it launched a paying crew of Italian air force personnel into suborbital space, apparently without incident, and has made several flights since. Virgin Galactic's spaceship is different from both SpaceX and Blue Origin in that its SpaceShipTwo craft (dubbed VSS *Unity*) takes off from a twin-fuselage aircraft (VMS *Eve*) while in flight rather than being launched from the ground via rocket. When *Unity* is attached to VMS *Eve*, *Eve* actually looks like an airplane with three fuselages. Recently, Virgin Galactic announced a hiatus in its tourist flights in order to transition operations to what it called a second-generation space vehicle. While this isn't necessarily a corporate death rattle, investors have become wary of the company's stock. Stay tuned.

A Few Reservations

The possibilities of commercial operations in space, particularly communications, global positioning, and off-Earth mining, are huge. The biggest source of payloads for the new generation of rocket builders will be the "constellations"—or "mega-constellations"—of low-Earth orbit communication satellites that companies like SpaceX, Amazon, Hughes Network Systems, and OneWeb are developing for deployment. The problem for rocket manufacturers, though, is that once these fleets of machines are deployed, demand for space transportation services will likely fall off sharply, at least for the short term. Like numerous automobile manufacturers of the previous century—Nash and Packard, Studebaker, the Briscoe Motor Company—some of these companies will fail.

Intriguing and potentially lucrative as it is, there are drawbacks to all this commercial activity as well. Pollution, for example. Many rockets burn nasty chemicals for propulsion. A favored fuel of the Soviet space program was UDMH, unsymmetrical dimethylhydrazine, combined with something called "red fuming nitric acid." Referred to as "devil's venom" by Russian scientists, this liquid propellant is—surprise!—highly carcinogenic to humans. Small amounts released at launch and the early stages of flight rained down on areas of Kazakhstan in the early days of Soviet rocket development and are said to have ruined sizeable amounts of cropland.

Rocket propellant-1, a highly refined form of kerosene used to fuel a number of modern rockets, including the SpaceX Falcon 9, is less virulent but still a potent source of black carbon pollution. Rockets using solid fuels spew tiny alumina particles, which may contribute to exhaust-related climate change. And it's not just exhaust-related contaminants we need to worry about. At least one U.S. satellite released radioactive plutonium when it reentered Earth's atmosphere in the sixties, and a Russian satellite called Kosmos-954 spewed radioactive material across Canada when it crashed in 1978. Though the fossil-fuel exhaust load of rocket flights is small compared to the aggregate filth produced by the millions of automobiles and tens of thousands of airplanes in the world, it's not negligible. These harmful effects will only get worse as the number of rocket launches increases—as commercial carriers like Blue Origin and SpaceX have promised they will. At least two companies, BluShift and Orbex, are trying to develop environmentally friendly rockets in response. The Orbex rocket is designed to run on bio-propane, a fuel produced from renewable feedstocks such as plant and vegetable waste material.

While it's not an immediate threat to Earth, "space junk" is a significant environmental hazard we've managed to create just above the planet. There are thousands of active satellites in Earth orbit. In addition to this functioning machinery, there are hundreds of thousands of pieces of orbital debris floating around Earth consisting of everything from entire spent rocket stages and derelict satellites (Telstar 1, for example, and SpaceX's so-called RatSat) to paint flakes, slag from solid rocket motors, and material associated with Soviet nuclear-powered reconnaissance satellites. The United States, Russia, China, and India have all contributed to the problem by testing anti-satellite missiles that blew up their targets and left pieces of the satellites whizzing around in orbit. Even seemingly innocuous items—like the beer bottle, Coke can, and roll of duct tape lost from the payload of one shuttle mission, courtesy of launch-pad workers who left the items in the orbiter's payload bay—are dangerous once they become part of the great maelstrom of miscellaneous objects clogging the heavens. Though many of these items are tiny, they're moving fast, and they're capable of causing considerable damage on impact. Along with naturally occurring micrometeoroids, human-manufactured debris poses a threat not only to active satellites but more importantly to crewed space vehicles and stations. Rogue paint flakes, for example, are blamed for cracking windows on space shuttle orbiters and the International Space Station.

Every new item added to the Earth-orbit freeway increases the odds of collision with other traffic, with costly and possibly tragic results. Indeed, in a so-far-theoretical scenario called the Kessler effect, low-Earth orbit could one day become so full of junk that a single collision could generate enough debris to cause a cascade of *other* collisions, creating additional debris, that would make space operations prohibitively hazardous for years to come. It is already fairly commonplace for astronauts on the ISS to have to shelter in place when threatened with such a collision. In fact, the ISS has had to undertake evasive maneuvers dozens of times since 1999 to avoid being hit by hunks of space garbage. And the first order of business for Chinese taikonauts on a recent mission to the *Tiangong* space station was to do a space walk to repair a solar array damaged by flying space junk.

Figuring out how to prevent the creation of additional space debris, and how to clean up what's already out there, will be increasingly pressing issues. China has been a serial space polluter, but it has also taken some early steps to address the problem of in-orbit collisions. In 2020 a Chinese satellite rendezvoused with a defunct satellite and towed it into an orbital path out of harm's way. In 2022 the Chinese demonstrated a "drag sail" that could be used to slow a derelict space object's speed and thus cause it to de-orbit and be destroyed by atmospheric reentry more quickly. In October of 2023 the U.S. Federal Communications Commission fined Dish Network, an American company, $150,000 for abandoning its Echo-7 direct broadcast satellite in an unacceptably low orbit, the first time any company has been disciplined for contributing to the space junk problem.

A second issue related to our growing skyfill (the space equivalent of landfill), more urgent for astronomers and other scientists than for the rest of us, is that satellite traffic is making it difficult for stargazers to study the heavens. Already the Hubble Space Telescope, which orbits at around 335 miles above Earth, frequently picks up images of satellites while attempting to view objects in deep space. The problem may be worse for terrestrial observatories. Investments in new astronomy facilities are being threatened by the degradation in viewing conditions caused by reflected light from these new artificial "constellations." And it won't be getting better any time soon. According to filings made with the FCC and the International Telecommunication Union, there are plans by various actors, both private and public, to launch some 431,713 satellites in coming years. The *actual* number likely won't be as high. But in combination with the thousands of satellites already deployed, it may be high enough to significantly degrade the quality of deep space imaging and observation.

Just as on Earth, progress in space will come with costs. We are just starting to figure out who will pay them.

Eleven Everyday Benefits
of the American Space Program

1. Satellite communications, including television, radio, telephone, and broadband internet services: Global connectivity is something we more or less take for granted today, but it all started with NASA's (and the Soviet Union's) first satellite launches, and the Cold War race for the moon.

2. Global Positioning System (GPS) data: Another huge satellite-related benefit, the GPS system pinpoints and tracks user locations all over the world. The system originated in a Department of Defense project called NavStar. The program's data became available to the public on a limited basis in the late eighties and more broadly beginning in 2000. Operated by the U.S. Space Force, GPS is now a pervasive feature of everyday life.

3. Weather satellite imaging: This is another space-related benefit we sometimes forget about. Orbital imaging provides us with warnings of large-scale weather events—most notably, hurricanes—that have saved countless lives.

4. Landsat imagery for use in mapping and environmental studies: Photographs from the original Landsat in the seventies were just the beginning. Follow-ups from *Skylab* and from subsequent Landsat missions and a slew of earth science satellites since have brought us huge advances in our understanding of Earth and its life support systems. The JPL website has a good current list of such missions and their aims, which continue to help us understand and care for our planet. The technology and the space-based data are there, folks. Whether we act in time to make a difference is up to us.

5. Microalgae supplements: Developed by scientists working to fortify food for the astronauts, a nutritional supplement made from microalgae and marketed as Formulaid is now contained in most baby food products sold in the United States and in many countries around the world.

6. LASIK technology: This now-common corrective eye-surgery method uses a laser-radar technique for tracking eye movement that was originally developed by NASA for use in autonomous rendezvous and docking of space vehicles to service satellites.

7. Water-filtering technology: Obviously, water filters didn't start with NASA. But the space agency had a strong incentive for getting the technology right, as astronauts on the ISS need to recycle and ingest their own urine in order to stay hydrated. NASA's work is now incorporated in many commercial products.

8. Memory foam: In the midsixties, NASA developed a material that was both soft and super shock-absorbent to help protect pilots in the event of a crash. This substance, an open-cell, polyurethane-silicon plastic that we now know as memory or "temper" foam, not only cushions seats for impact but also increases their comfort. It's now found in all sorts of products, including car seats, bike seats, military gear, and, of course, mattresses.

9. Infrared thermometers: Because there's no way to reach out and touch them, NASA has used infrared technology to measure the temperatures of stars. Private industry borrowed the idea on a more modest scale to measure the temperature of human beings, which is especially useful when a caregiver is attempting to minimize close contact with a patient.

10. Lightweight breathing systems for firefighters: NASA research has resulted in a number of advances useful in both firefighting and fire prevention: protective outer garments for workers in hazardous environments, a broad range of fire-retardant paints and foams, fire-blocking coatings for outdoor structures, variable-sensitivity smoke detectors, and a number of flame-resistant fabrics for use in the home, office, and public transportation vehicles. Among the most impor-

tant NASA-related advances is the creation of portable breathing systems that shaved weight and size from existing models, making the gear easier to carry and the firefighter more mobile.

11. "The Grey," episode 10 of season 2 of *For All Mankind*: Hey, entertainment is important too!

18

Curse You, Gene Roddenberry!

Science fiction, including producer Gene Roddenberry's influential
Star Trek TV series, makes interstellar travel look easy. But the fact is,
human beings aren't going anywhere beyond Jupiter without some significant
advancements in technology—especially propulsion technology.

We've been to the moon, and we plan to return—this time, perhaps, to stay. And an American, Chinese, or multinational group of Earthlings will almost certainly visit Mars within the next century. But getting elsewhere in our solar system, much less *outside* the solar system, is another matter. How will we hydrate and feed our spacefarers on a multiyear trip to and from one of Jupiter's moons? How can we shield them from cosmic radiation on their long journey and once they arrive? How will we keep our astronauts physically fit and psychologically centered as they speed away from everything they've ever known into a future of question marks, cold, and infinite darkness?

These are all good questions. They deserve close study. But the biggest issue we face in sending human beings to distant planets isn't related to food, fitness, or radiation. The biggest issue is figuring out how to get there. Given the current state of rocket propulsion, the answer is, *We can't*. It is currently impossible to send people to planets beyond our solar system.

Live people, anyway.

Consider the numbers. Light travels at a velocity of 186,282 miles per second. That's faster than anything we know of. It's apparently as fast as anything *can* travel. To express the distances between objects in the universe in a way that doesn't rapidly devolve into long sequences of zeros, we speak in terms of the distance light travels in the course of a year. This distance, one light year, is 5.88 *trillion* miles. The distance from Earth to Proxima Centauri, our nearest neighboring star, is around 4.25 light years, or 25,300,000,000,000 miles. By contrast, the distance from the sun to Earth is only 93,000,000 miles, or around 499 light *seconds*. This means that light from the sun reaches us 499 seconds, or just over eight minutes, after it leaves the solar surface. The distance from the earth to the moon is a mere 240,000 miles, or roughly 1.3 light seconds.

Keeping these numbers in mind, note that it would take approximately thirty-nine years traveling at the speed of light to reach the Trappist-1 solar system, which is believed to contain several Earthlike planets. This would almost be imaginable if we could travel at the speed of light. Unfortunately, we can't. We can't even get close. According to Space.com, if a spacecraft were to travel at the same speed as seventies-era space probe *Voyager 1*—around 38,200 mph—it would take roughly 685,000 years to get to Trappist-1. Proxima Centauri appears to have a planet in its orbit that might be like Earth. This would be exciting news, except that at *Voyager*-type speeds, it would take us *seventy-three thousand years* to arrive at this relatively proximate planet. That's a long time for hundreds of generations of crews to spend drinking recycled urine and watching *Better Call Saul*.

In science fiction, we've been journeying back and forth across the cosmos for decades. In the *Star Trek* universe, we can use warp drive (as long as the dilithium crystals hold out!). We've jumped into "hyperspace" in a galaxy menaced by Darth Vader, activated the Epstein drive in *The Expanse*, and, in the brain-bending 2014 movie *Interstellar*, plunged into a trans-galactic wormhole shaped like a snow globe in hopes of finding habitable worlds somewhere Out There. None of these techniques has ever been created or attempted in real life, of course, which means that we're stuck with what we know. And this is a problem, because even a trip to Mars, our most hospitable solar system neighbor, will take something like eight months using current propulsion systems. Chemically fueled rockets like the Saturn V, the SpaceX Falcon 9, and the ESA's Ariane 5, typically using a combination of liquid oxygen as an oxidizer and either rocket-grade kerosene or liquid hydrogen as fuel, have been the powerhouses of the space age. They are still the only type of vehicle that can generate enough thrust to lift people and payloads up out of Earth's gravity well and into orbit. But the propulsive power of these rockets is fundamentally limited by the energy held in chemical bonds. The relative efficiency of the thrust they generate—what engineers call their *specific impulse*—just isn't going to get much better. Furthermore, carrying these fuels is itself a problem, as they are difficult to maintain and add substantial amounts of mass to a spacecraft.

Some scientists therefore argue that we should be figuring out how to generate propellants on other planets, rather than carrying them with us. For example, crewmembers of a mission to Mars might split ice obtained from the Martian poles into hydrogen and oxygen to use as rocket fuel for the trip back to Earth, or on to other planets. But others contend that chemical rocketry will *never* be enough to get us where we want to go in any reasonable time frame. They believe we need to explore—or in some cases, *resume exploring*—the use of alternative fuels and propulsion methods that will allow us to travel farther and faster once we're out of Earth's gravitational grip. The best known and most promising of these alternative propulsion methods involves nuclear fission.

NASA and the Defense Advanced Research Projects Agency (DARPA) announced in January 2023 that they would work together to develop a nuclear thermal-powered rocket by 2027. The project, called Demonstration Rocket for Agile Cislunar Operations, or DRACO, will allow astronauts to "journey to and from deep space faster than ever—a major capability to prepare for crewed missions to Mars," according to NASA administrator Bill Nelson. Faster in this case means three or four months to Mars rather than eight. It's still a grueling trip, but the prospect of reducing travel time by 50 percent is enticing. And it can be done. We've known for decades how to power naval vessels and satellites through nuclear fission. It can work for crewed spaceships as well. The European Space Agency also recently announced plans to fund new research on nuclear propulsion for use in deep space exploration.

The alternative propulsion possibilities don't stop with fission. The dream of nuclear *fusion*, the combination of atoms rather than the breaking of them, is very much alive, if still elusive. If we could master the technical and material challenges of slamming hydrogen isotopes together at the hellishly high temperatures required to generate energy, we could effectively travel on the strength of tiny stars carried in our spaceships. Carried in massively reinforced tungsten *vaults*, admittedly, but you get the idea. Such travel would require relatively modest amounts of hydrogen for fuel, and this fuel would be essentially inexhaustible, since hydrogen is by far the most plentiful element in the universe. Hydrogen is basically everywhere. In the seas. On the moon. If the universe is a bakery, hydrogen is its flour—the oldest, simplest, most common stuff in creation. Fans of the speculative fiction television series *For All*

Mankind will note that all three of the spaceships racing to Mars in season 3 are nuclear-fusion powered, though (*Spoiler Alert! Look Away!*) one of them also carries solar sails to speed it on its way.

Alas, fusion is so far only a pipe dream. Every few years, a "breakthrough" is announced, but there always seems to be a caveat. Fusion has been demonstrated but not dialed in. It remains propulsion's holy grail, tantalizing and elusive.

Other propulsion prospects for deep space journeys involve ion thrusters, used to date on a number of satellites and on NASA's *Dawn* probe; matter/anti-matter reactions, theoretically super-efficient at producing energy but so far wildly impractical; and the aforementioned solar sails. Particles of light, called photons, carry tiny but measurable amounts of momentum. The notion behind using a "solar sail" is to catch enough of the sun's photons to push a spaceship, in much the same way that wind moves a sailing ship at sea. Solar sails provide low levels of thrust because the energy captured by the sails is relatively small. But while the energy harvest is small, it's also constant, which means that acceleration increases as long as the sails receive sunlight. The result is that solar sail-powered spacecraft can eventually accelerate to impressive speeds. A significant problem for solar sail enthusiasts is that "photon power" grows weaker as one travels away from the sun. This makes sails impractical much beyond Mars and more or less useless in interstellar space. One alternative: using a giant laser to fill the sail instead. This method of propulsion has been proposed for a probe to Proxima Centauri. Scientists associated with a project called Breakthrough Starshot have theorized that with a large enough sail, a powerful enough laser to fill it, and a small enough— like, *postage-stamp size*—spacecraft, an interstellar probe could be accelerated to a speed of one hundred million miles per hour, fast enough to reach Proxima Centauri in twenty years.

There's at least one other, even more exotic, space travel notion. Warp drive via Alcubierre engine, anyone? As best we can understand it, this fantastically theoretical means of propulsion involves creating "negative mass" to contract certain areas of space while enlarging others and riding the resultant "wave" or warp of space to travel faster than the speed of light. Maybe the better way to say it is not that we would be traveling faster than light, which is, again, apparently impossible. Rather, we would be shrinking the distance we would need to travel, thus *effectively* traveling faster than light, though not actually doing so.

Ready for a CBD gummy?

The fact is, we're stuck. Humanity is standing at the edge of a canyon. We're reaching out as far as we can, but the light of distant stars and the promise of all their brilliant worlds are still well beyond our grasp. We need a technological leap just to get us to the point where we can visit Mars and return in a reasonable amount of time. Voyaging outside of our solar system will require us to generate energy in amounts many orders of magnitude larger than what we are capable of creating today. Bottom line: Look for the first successful Mars mission to be nuclear-thermal powered. And don't count on human beings venturing much farther than Enceladus unless and until some more efficient and powerful propulsion source becomes available.

We're confident that it will, by the way. People laughed at Robert Goddard too. But now, a hundred years later, nobody laughs at a chance to visit the International Space Station. There are two brands of humbuggery when it comes to the prospects for space travel. One is unbridled optimism. The other is It'll Never Happen-ism. The truth is somewhere in the middle. Maybe the most important takeaway from any discussion of propulsion problems is that the scientists and engineers working on solutions need financial support. Without taxpayer dollars and significant amounts of public and congressional encouragement, it's going to take them (and us) a heck of a lot longer to figure out how to explore our solar system and beyond.

Rockets, it turn out, run mostly on money.

The Eleven Most Persistent Rumors, Riddles, and Conspiracy Theories about the American Space Program

1. We never landed on the moon: The Big Daddy of them all. Oddly persistent. And like the best (and worst) conspiracy theories, capable of hosting numerous contradictions beneath its gown. But if you believe that the four hundred thousand Americans who worked to make the moon landing happen could all keep a secret this big, we've got some crypto-currency to sell you. Bonus points for explaining why the Soviet Union, sworn rival and longtime detractor of the American space program, has never claimed that the moon landings were staged.

2. Sure, you can land, but . . . : This theory posits that *Apollo 10*'s lunar module was given only half a tank of fuel to ensure that the men flying it—astronauts Tom Stafford and Gene Cernan—wouldn't try to land on the moon after they made a scheduled "dress rehearsal" approach to within eight nautical miles of the lunar surface. It's not true. The threat of a punch-up at the hands of Deke Slayton was enough to forestall any such possibility.

3. Neil Armstrong was a robot: We made this one up. It's not true. Probably.

4. If worse comes to worst: Some say the NASA astronauts were given cyanide tablets so they could kill themselves in case of an in-space emergency, saving themselves and the world the agony of watching a slow death by hypoxia or hypothermia. *Apollo 13* astronaut Jim Lovell says it's not true, and he should know.

5. Someone has to do it: Tom Stafford was told that if Gene Cernan was unable to return to the capsule after his scheduled space walk

on *Gemini 9*, Stafford should cut him loose and leave him in space. *True*. Horrifying, but necessary.

6. It's for your own good: Some sources claim that the space shuttle orbiter's hatch was locked to prevent astronauts from attempting to open it during orbit. *True*. NASA evidently adopted this safeguard after a payload specialist astronaut grew so despondent over the failure of a machine he needed for an important experiment that he threatened to, um, *not return to Earth*. This is not the sort of talk you want to hear at 150 miles above Zambia, where opening a hatch would result in immediate depressurization of the orbiter and the rapid deaths of all aboard. Hence the lock.

7. Wipe that moondust off your visor, mister: Rumor has it that both the army and the air force hatched plans for fortified lunar bases in the early sixties, complete with working moon guns and cosmic bazookas. *True*, actually. The plans were scuttled when the services couldn't find anyone to work in the PX.

8. The boss says go: Some observers speculate that the fateful decision to launch *Challenger* on 28 January 1986 occurred because NASA officials desperately wanted to please President Ronald Reagan, who was scheduled to give his State of the Union speech that evening. It's true that *Challenger* carried America's Teacher, Christa McAuliffe, and was of interest to the president. He would certainly have played up the flight in his speech. But while getting *Challenger* up in time for the State of the Union address may well have been a consideration, there's no direct evidence of the link. NASA had developed a bad case of hubris by this point anyway, compounded by the comorbidity of Go Fever, and the decision to launch on that frosty Florida morning might well have been made regardless of Reagan's speech.

9. There is a door on Mars that leads to ruins of a mysterious Martian civilization: NASA's *Curiosity* rover did indeed find a door-like cleft in the rocks of the Red Planet's Gale Crater. However, the "door" is quite small and can doubtless be explained by natural causes—as can other features previously identified, such as the "face" found by *Viking 1* in 1976, since revealed to be a trick of shadows and light rather than a monument to some long-vanished race of solemn Martians.

10. NASA considered building a rocket powered by nuclear explosions: This one's too strange to make up. It's true—and fascinating. Read about Project Orion in Rod Pyle's great book *Amazing Stories of the Space Age: True Tales of Nazis in Orbit, Soldiers on the Moon, Orphaned Martian Robots, and Other Fascinating Accounts from the Annals of Spaceflight.*

11. Saturn is wearing a hexagonal, multicolored yarmulke: Again, too bizarre for us to come up with, and absolutely true. The ringed planet, one of the solar system's two gas giants, has a stunning north polar vortex with a distinctively symmetrical shape. First discovered by astronomer David Godfrey on the basis of imagery captured by NASA's Voyager program, scientists speculate that the feature is caused by the intersection of winds traveling at different velocities. On the other hand, it might have been created by Neil Armstrong. Who may have been a robot.

19

Looking Outward

Mathematics tells us that we almost certainly have company in the universe—
maybe even *lots* of company. But if that's the case, where is everybody?

The Hubble Space Telescope is over three decades old, well beyond its initial life expectancy. The occasionally temperamental satellite appears to be nearing the end of its operational life, which will mean we have to say goodbye to the rush of receiving all those bizarrely beautiful galactic snapshots it's been sending us over the years. But all is not lost. There are other space telescopes either currently on the job—the Chandra X-Ray Observatory, for example, and the Transiting Exoplanet Survey Satellite (TESS, otherwise known as Explorer 95)—or being prepared for duty, like NASA's SPHEREX near-infrared space observatory. The most promising of the new generation of celestial exploration devices is undoubtedly the James Webb Space Telescope (JWST), which is sometimes called the new Hubble and has at other times been referred to as the "telescope that ate astronomy," due to its high cost and scheduling delays. But that was then, during the seemingly interminable delays before JWST was actually deployed. No one's complaining now. JWST's images are even sharper than Hubble's. We are living in a golden age of astronomy, an era whose discoveries are all the more astonishing for the speed with which they accumulate.

JWST was launched from South America atop a European Space Agency Ariane rocket in December of 2021. On its way to establishing a stable orbit almost a million miles from Earth, the telescope went through an arduous and complicated deployment sequence—a sort of in-transit "unfolding," or reverse origami—without a hitch. NASA released the telescope's first photos to the public on 12 July 2022. JWST isn't exactly a replacement for Hubble, since it "sees" in the infrared range of electromagnetic wavelengths exclusively, while Hubble is receptive to a wider range of signals. But JWST's mirrors are significantly larger than Hubble's. And because the new observatory is farther away from the sun and protected from our star's radiation by a solar shield, JWST stays cooler, which is crucial for detecting infrared radiation. The upshot is that the new telescope is able to peer much deeper into the universe. One problem with this remote placement: JWST is essentially impossible to service. If it fails or is damaged, there's not a lot we can do about it. We should enjoy it while we can. JWST has already been used to study the most distant galaxies in the universe, and the atmosphere of an exoplanet (a planet outside our solar system) seven hundred light years from Earth. It promises to rewrite the astronomy books that Hubble just finished revising. Hubble got us into the theater. JWST is opening the curtain.

It's not unusual to read claims that Hubble and JWST are able to "look back in time." It's a catchy phrase, but the reality is less exciting than it sounds. The thing to remember is that everything we see in space has already happened. This is because even light, which is the fastest thing we know of, takes time to travel. Light from the sun, for example, takes eight minutes and twenty seconds to reach Earth after it leaves its source. Thus, we on Earth are always seeing the sun as it existed a little over eight minutes earlier. When astronomers peer at stars that are ten light years away, as they can with Hubble, JWST, and other instruments, they are seeing and recording images of the stars as they existed ten years ago. Those stars may have changed since then. In fact, stars that the Hubble Space Telescope has observed *millions* of light years away may not even exist anymore. What we're seeing is what they looked like at a period deep in their history. That's what journalists mean when they breathlessly report that we are now "looking back in time." It's exciting, yes. But we haven't exactly stepped into the TARDIS, Dr. Who's inter-dimensional Uber, capable of visiting past and future centuries and famously larger on the inside than on the out.

The next major deep-space observation satellite to take to the skies will be NASA's Nancy Grace Roman Space Telescope (the RST). The instrument is named for NASA's first chief astronomer, who was also the agency's first female senior executive and is sometimes called the Mother of Hubble. RST is currently scheduled for launch in 2027. Like Hubble, the satellite will gather infrared energy from distant space. And while RST's mirror will produce images of similar quality, RST will be able to survey sections of the galaxy approximately a hundred times wider than Hubble. Astronomers hope RST's wide scans will help them find objects of interest, which can then be studied more closely with JWST's more powerful equipment. Another of RST's functions will be to map what NASA calls "enormous panoramas of the universe" in order to study so-called dark energy, the mysterious force that scientists believe is pushing our universe to expand. The new telescope will survey millions of galaxies, mapping our cosmological neighborhood to give us an idea of how the universe has evolved. RST will also attempt to use gravitational microlensing—tiny changes, or "curving," in starlight—to discover a multitude of exoplanets, and it will house a coronagraph that will use new starlight-suppression technology to allow astronomers to study such planets.

Despite the advantages and considerable success of Hubble, JWST, and other space telescopes, Earth-based observatories continue to provide important observations. The most exciting terrestrial astronomy project currently in view is the Vera C. Rubin Observatory. This isn't a NASA project—it's funded, at least in part, by the U.S. Department of Energy—but it's certainly a part of the American space exploration program writ large and worth noting for anyone interested in astronomy generally. Astronomers refer to the initial operations of a telescope as its "first light." Vera C. Rubin's first light is scheduled for late 2024. Perched in the highlands of northern Chile, the telescope will eventually deliver some five hundred *petabytes* of images from our cosmic environment. According to the observatory's website, the aim of the observatory is to conduct a deep survey (called the Legacy Survey of Space and Time, or LSST) over an enormous area of sky, do it with a frequency that enables images of every part of the visible sky to be obtained every few nights, and continue in this mode for ten years to achieve astronomical catalogues thousands of times larger than have ever previously been compiled. While Hubble has provided snapshots, Vera C. Rubin will paint a huge mural, employing, among other tools, an 8.4-meter-diameter main mirror and the largest digital camera ever assembled. Former astronaut and current meteor fighter Ed Lu calls the new observatory the "big dog" that will enable unprecedented mapping of potentially dangerous asteroids—a crucial enterprise for protecting the earth.

Major Minor Missions

To the extent the American media reports on space at all, the big boys—NASA and SpaceX, Bezos and Musk, Starliner and the ISS—get most of the headlines. That's understandable. Square-jawed rocket riders, grinning space tourists, and sunrise launches stir the imagination. And NASA's plan to get us back to the moon by way of the Artemis program is exciting, though it's certainly taking longer than initially advertised. Lost in the prop wash of these splashy crewed missions are some fascinating—and occasionally quirky—science expeditions.

Of course, any flight into space is to some extent a "science" mission, since we're still learning on the job. But these sojourns are different. They're all about innovation, surprise, and disruption. They bear their names like flags: *Opportunity*, *Curiosity*, and *Spirit*; *Kepler*, *Cassini*, *Huygens*, and *Galileo*. The *Perseverance*-rover-and-*Ingenuity*-helicopter buddy movie that played out on Mars from 2021 until 2024 is a good example. So were NASA's 2021 probe of the carbonaceous near-Earth asteroid 101955 Bennu to pick up a few soil samples, and its *Dawn* ion thruster–powered mission to the dwarf planet Ceres and the asteroid Vesta in 2015. Perhaps the most dramatic demonstration of science for science's sake was NASA's 2015 flyby of Pluto. The agency's *New Horizons* probe, around the size of a piano, according to NASA, traveled nine years and three billion miles to reach Pluto's home in the icy Kuiper belt. Once there, the *New Horizons* probe captured jaw-dropping images of what astronomers realized was a tectonically active, and possibly *volcanic*, orb. Even Pluto, distant and demoted, harbors mysteries for interested observers.

Perhaps the most important of NASA's lower-profile missions is NASA's Double Asteroid Redirection Test, or DART. An asteroid is a naturally occurring space object that is smaller than a planet (though there's some gray area here) and larger than, say, a microwave oven. Anything smaller is a meteoroid. There are untold numbers of such objects in our galaxy and many millions in our solar system alone. Given the vast numbers involved, we encounter them (in other words, *collide* with them) fairly rarely. But when it happens, the results can be dramatic. Asteroids and meteoroids become meteors when they enter Earth's atmosphere, showing up in the sky as bright streaks of light. The small ones burn up before they hit the ground, and their brief flicker of flame is often held to be beautiful and fortuitous. People write songs about these "falling stars." But large asteroids—and thus large meteors—pack enough kinetic punch to wipe out huge portions of our planet's flora and fauna. The so-called Chicxulub Impactor, a six-mile-wide slab of rock and metal that slammed into the Gulf of Mexico some sixty-six million years ago, changed the pace and direction of animal evolution by wiping out much of the planet's dinosaur population. Even smaller asteroids could easily destroy cities and kill millions of people. While NASA currently knows of no planet killers heading toward us, it's only a matter of time before one presents itself. Meanwhile, smaller space bombs approach us every day. A meteor over the Russian city of Chelyabinsk arrived unannounced in February of 2013. It exploded at an altitude of around eighteen miles, generating thirty times as much energy as the atomic bomb dropped on Hiroshima and injuring some 1,500 people.

Enter DART. Or rather, *depart* DART. DART left Earth on a Falcon 9 rocket on 24 November 2021 to begin a seven-million-mile journey toward the binary asteroid system Didymos. A binary system means that there are two distinct asteroids orbiting each other as they travel through space. On 26 September 2022 DART slammed into the smaller of the system's two asteroids—Dimorphos, an asteroid "moonlet"—at fourteen thousand miles per hour. The mission bore some resemblance to the Deep Impact project of 2005, when NASA sent a probe hurtling into the comet Tempel-1 in an attempt to learn what lay beneath the surface of that object. The point of DART, though, was to see if Dimorphos's orbit could be altered as a result of the impact, and indeed it was. What was altered in the experiment was an orbit, not a course toward Earth, but the implication is clear: we can change the path of an asteroid by firing a projectile at it from a long distance away. And while the effect of impact on Dimorphos was relatively small, even a slight nudge and resultant course alteration over the span of several million miles could be enough to steer an asteroid headed for Earth harmlessly off into some other sector of the cosmos. And this is important, because the numbers are chilling. Seventeen meteors fall to Earth each day. Experts and scientific researchers have stated that in the past six hundred million years, some sixty asteroids or similar objects of three miles or more in diameter have hit our home, and three impact events are thought to have caused mass extinctions of plant and animal life. In light of such facts, it seems prudent to know how to defend the planet. DART is a promising step in this direction.

Another intriguing science mission was launched from Cape Canaveral Space Force Station in October 2021, when the probe NASA calls *Lucy* set off to study Trojan asteroids. Launched atop a ULA Atlas V rocket, *Lucy* was placed in a heliocentric orbit, then used Earth's gravity to sling herself out toward her first destination on a twelve-year mission. Most of the millions of asteroids in our solar system are of no particular interest. Among them, though, is a subset that scientists are keen to study. They are the Trojan asteroids, space objects that are caught in the competing gravitational pulls of the sun and a planet and which consequently share the orbital path of that planet, locked in a more or less permanent space race in which the positions never change. So far we've seen these sorts of asteroids only in our own solar system, though they presumably exist in others as well.

The first Trojan asteroid was observed in 1906 near the planet Jupiter. We now know that Jupiter has many thousands of Trojans. In fact, there are two groups, or "swarms," of them. One occupies space about sixty degrees in "front" of Jupiter's orbital path, and the other group moves through space about sixty degrees behind the giant planet.

Asteroids are a subject of interest these days at least in part because they may contain minerals that are valuable here on Earth. Famously, one asteroid, 16 Psyche, is suspected to be made of iron and nickel in amounts that could be worth trillions of dollars on Earth. No such claims are made at present for Jupiter's Trojans. Nevertheless, some may contain ice that could be harvested for consumption or commercial use in space. And because astronomers believe that these Trojans were trapped in their orbits at or near the formation of our solar system, all of them are artifacts of deep time that may be able to help us understand how our local planetary neighborhood was platted and what it's made of. As NASA puts it, Jupiter's Trojans are "time capsules from the birth of our Solar System more than 4 billion years ago, [and] thought to be remnants of the primordial material that formed the outer planets."

It is the notion of Jupiter's Trojans as developmental relics that gave the mission its name. After all, Lucy is the moniker given to the collection of bone fragments found by paleontologist Donald Johanson in Ethiopia's Afar Valley in 1974. This tiny constellation of splintered calcium indicates that proto-humans like Lucy probably learned to walk upright before the rapid expansion of brain size that eventually made us human—an important clue in understanding evolution. Not coincidentally, the name of the "main belt" asteroid that *Lucy* will visit in 2025 is 52246 Donaldjohanson. Fossils on Earth, fossils in space. It's a neat conjunction.

Another little quirk of the *Lucy* mission: the probe carries a sort of greeting card bearing quotes from notable Earthlings, including all four Beatles and Yoko Ono, Martin Luther King Jr., the poet Amanda Gorman, and secular prophet Carl Sagan. The thoughts are harmless enough, and it's an amusing callback to the plaques and golden records sent into interstellar space in the seventies on the *Pioneer* and *Voyager* missions, when we were eager to post Tinder profiles of ourselves for all the galaxy to see. But since *Lucy* will almost certainly never leave our solar system, it's unclear who the messages are meant to reach. Future Earthlings, perhaps? Or settlers on one or more of Jupiter's moons?

No matter. *Lucy* has set out for the future to tell us about our past.

The Truth Is Out There

The prospect of finding life on other planets has intrigued and unnerved humanity for a long time. It's a preoccupation of science fiction, on paper and in film, from H. G. Wells's *The War of the Worlds* through the *Aliens* movie franchise and more recently in Andy Weir's wry and inventive novel *Project Hail Mary*. Indeed, it would be difficult to count the various extraterrestrial races—Klingons and Cardassians, Harvesters and xenomorphs, Kanamits, the Borg, the Kree and the Blob—invented by novelists, artists, and film-makers over the years.

And we do mean *years*. Fascination with the possibility of alien life dates back at least four centuries. Writers in the seventeenth century, for example, posited the existence of intelligent societies on Mars and Venus. A famous New York newspaper hoax in 1835 involved "reporting" on the telescopic discovery of winged humanoids living on the moon; the paper's circulation soared as a result. Orson Welles's radio broadcast of *The War of the Worlds* a century later produced a similar, though shorter-lived, hysteria. The 1950s and early '60s brought cinematic invasions by the likes of triffids, pod people, and the shape-shifting Thing. In the late '60s and '70s, humans met up with more helpful alien races in *Star Trek* and *E.T. the Extraterrestrial*. Then things turned dark again. Sigourney Weaver's Ripley battled nasty alien insects with a taste for human flesh, and Arnold Schwarzenegger went hand-to-hand (*hand-to-claw?*) with a dreadlocked interstellar predator that traveled to Earth to hunt people for sport. In James Cameron's *Avatar* movies, *human beings* became the alien invaders, with devastating consequences for the flora and fauna of a distant world but—let's face it—amazing merchandising opportunities.

It's not just creative types who have tried to figure out if life exists elsewhere in the universe, and, if so, whether it would eat us. An increasing number of astronomers, biologists, and other scientists are studying these questions as well. As an article on Science.com wonders:

What might we find: little green men or microbes? How might we find them: radio waves or strange chemicals in [a] planet's atmosphere? Something no one has even thought of yet? Over the decades, scientists considering the possibility of life beyond Earth have pondered what such life might look like, how humans might be able to identify it from afar—and whether communication between the two worlds might be possible. That thinking has included developing classification systems ready to fill with aliens. One such system is called the Kardashev scale, after the Soviet astronomer who proposed it in 1964, and evaluates alien civilizations based on the energy they can harness.

Nikolai Kardashev proposed his scale in a paper titled "Transmission of Information by Extraterrestrial Civilizations," one in a number of theoretical ruminations he published about the possible capabilities of life elsewhere in the universe. It can fairly be asked, though, Why start *classifying* alien civilizations when we haven't found a single one yet?

For optimists, the answer is largely mathematical. In 1961 astronomer Frank Drake attempted to calculate the number of alien civilizations that might exist in our galaxy. Starting with the rate of star formation in our galaxy, he added a number of variables, such as the likelihood of such stars having planets and then of their having *habitable* planets, and he worked through this list of variables to arrive at an estimate or range of the number of planets in the Milky Way that might host intelligent civilizations. He arrived at a projection of somewhere between a thousand and a hundred million civilizations.

It's a stunning set of numbers, of course, though such calculations are more valuable as a spur to discussion than an actual "answer." Jon Gertner, writing for the *New York Times*, put it this way: "Because there are probably at least 100 billion stars in the Milky Way galaxy, and an estimated 100 billion galaxies in the universe, the potential candidates for life—as well as for civilizations that possess technology—may involve numbers almost too large to imagine." Indeed. The physicist Michio Kaku supposes that one in every two hundred stars has one or more Earth-like planets orbiting it. This means that in our galaxy alone, half a *billion* stars may have Earth-like planets in tow—a huge number of habitable, and possibly *inhabited*, worlds. Astronomer Avi Loeb goes further, speculating that some fifty billion worlds with "life-friendly conditions" exist in our galaxy. "And that," he adds, "is counting only habitable planets within the Milky Way. Adding all other galaxies in the observable volume of the universe increases the number of habitable planets to . . . 10 to the 21st power, a figure greater than the number of grains of sand on all the beaches on Earth."

Simply put, the strongest argument for the existence of extraterrestrial life is statistical. Regardless of whose calculations you use, the figures are difficult to ignore. Is humanity the lucky winner of the universe's one-in-a-sextillion golden lottery ticket? Is ours the only world that does the watusi? Are love and loss utterly unknown in the far alleys of existence? Sheer *numbers* argue for the possibility—some would say the probability—of life elsewhere. And then, of course, the imagining begins. Octopus-like creatures who live in the mist. Silicon-based beings clinging to rocks to bask in the warmth of a dying sun. Air-breathing jellyfish with super brains. Terrifying hairless bipeds who consume the flesh of other species and—*oh, wait.* That's us.

On the other hand, as physicist Enrico Fermi once put it, in what has become known as the Fermi paradox, if there are so many alien civilizations out there, why haven't we seen evidence of them? In other words, where *are* they? Grainy videos and breathless hearsay aside, evidence of alien visitation is nonexistent. This makes sense. Even assuming that intelligent life does exist elsewhere in the universe, there's a long list of reasons why we might not have encountered it yet. First, given the size of even our quiet little stellar neighborhood, any extraterrestrial civilization is likely to be mind-bogglingly far away and faced with the same problems getting across the vast distances of the universe that we are currently confronting. Second, we Earthlings are a fairly young species. Numerous cosmic federations, foundations, and empires might have existed, thrived, and crumbled into dust elsewhere in the universe many millions of years before humankind ever invented the electric guitar. Third, alien life might not consider us *worth* contacting, as any race of beings capable of observing us without our knowing about it would necessarily be more technologically advanced than we are. Why would they need electric guitars anymore?

Then, too, talking about the possibility of alien life involves the question of what exactly life *is.* Even here on Earth, we wrestle with whether certain forms of organized energy—viruses, for example—are "life." While we suspect that life elsewhere might well arise from the sort of environmental circumstances prevailing here on Earth, such as the presence of water and warmth and high concentrations of carbon, it's possible that life forms elsewhere could be entirely different. And even life akin to what we know could be capable of adapting to unlikely environments. "Extremophile" (literally, "extreme loving") organisms on this planet, for example, have proven capable of existing in thermal flumes at the bottom of the sea, under Arctic ice, and in nuclear reactors. Some terrestrial bacteria can even survive prolonged exposure to the vacuum of outer space. Why not, then, on one of Saturn's 146 moons? But even if some sort of life exists elsewhere, in or out of our solar system, it's another large leap to think it would evolve in such a way as to bring it to where we are now: waiting, and wondering, unsure of what exactly it is we're looking for out there, and what it will mean when we find it.

Or it finds us.

All along the Watchtower

There have been numerous efforts to locate evidence of life elsewhere in the cosmos, efforts commonly referred to under the acronym SETI, or the search for extraterrestrial intelligence. Most of these searches have focused on monitoring sections of the sky for radio signals. At one point NASA oversaw such monitoring, under the auspices of the High Resolution Microwave Survey. This program used several radio telescope facilities, including the giant Arecibo Telescope in Arecibo, Puerto Rico, to eavesdrop on the heavens. Funding for the survey attracted ridicule, however, and Congress gave it the axe in 1993, a year after it started.

Since then, the sky search has been carried on by universities, private organizations, and dogged individuals. One major supporter and advocate of the search is the California-based SETI Institute. ("Where will you be," SETI's website asks, "when we find life beyond Earth?") So far, however, after at least fifty years of not entirely coordinated but still serious searching, the scientific consensus is that there has been no contact. Perhaps the most tantalizing possible electromagnetic feeler is the so-called Wow! Signal picked up by researchers using Ohio State University's Big Ear radio telescope. On the evening of 15 August 1977, just one day before Elvis Presley died, the facility recorded a radio signal seemingly coming from the Sagittarius constellation. Lasting seventy-two seconds, the signal was confined to a narrow frequency and was unusually strong. Its detection was reflected in a computer printout that a young astronomer named Jerry Ehman reviewed. Seeing the sudden spike in signal intensity, he scrawled "WOW!" next to the read-out, which used the letters 6EQUJ5 to denote the signal's strength. This phenomenon has never been recorded since, which means that it was essentially meaningless for scientific purposes. Still, it has provided many astronomers, SETI enthusiasts, and Elvis fans considerable food for thought—and speculation.

The Transiting Exoplanet Survey Satellite, the James Webb Space Telescope, and other observatories and space telescopes are looking for spectrographic chemical signatures that would indicate the presence on exoplanets of conditions favorable to or reflecting the presence of life. Perhaps the most important biosignatures in this regard are oxygen, indicative of a planet with surface water, and methane, which is commonly produced as a waste product of carbon-based life forms. More recently, though, scientists have also begun looking for "techno-signatures," signs not of life itself but of life's creations—light, pollution, and atmospheric changes. A conclusive clue—more like a billboard, really—would be the presence of so-called Dyson spheres around distant suns. Dyson spheres are theorized mega-structures—in this case, mega-*mega*-structure might be the more appropriate term—that could be built to surround a star and, with suitable technology, harvest a large percentage of the star's energy for use by the civilization advanced enough to construct such a device. The sphere might also be used to reduce a planet's exposure to solar flares and other harmful radiation.

Yet another avenue of investigation is more cerebral. For example, NASA funds the Laboratory for Agnostic Biosignatures, a group of scientists trying to understand the ways in which alien life might not be like earthly life at all. They are, their website claims, "developing techniques to detect life in the universe that humans can't conceive of." Such beings might, for example, be silicon-based rather than carbon-based. Life elsewhere might evolve using genetic maps different from our own familiar DNA and RNA sequences. It might indeed be entirely microbial. The search for biology elsewhere in the universe is an exciting quest. In fact, it may be humanity's *most* exciting quest. But it should be tempered by the realization that even if we find indications of extraterrestrial life, it might take many thousands of years to interact with it. Whether that's a good thing or a bad thing depends of course on one's estimation of the nature of life elsewhere. It's possible that some of it simply wants to serve man.

Have We Been Visited?

The search for alien life is naturally related to an interest in alien life's possible search for *us*—that is to say, an interest in what is often referred to as UFOs, or unidentified flying objects. It is almost impossible to attend an event featuring an astronaut without hearing someone ask about UFOs. For many years, NASA scrupulously avoided this subject, which has long been regarded as the province of moon bats, mumblers, and conspiracy theorists. Maybe the biggest recent news about UFOs is the agency's announcement in June of 2022 that it would join the search for answers about "unexplained aerial phenomena," NASA's briefly preferred but essentially synonymous terminology. We say "briefly" because NASA has more recently stated that it actually prefers the term "unexplained *anomalous* phenomena," or "mysterious weird stuff," which actually isn't as good a term as its two predecessors because it doesn't even refer to airborne events. The agency might as well be investigating chupacabra sightings.

Nevertheless, NASA appointed a blue-ribbon panel to look into UAP data. "Exploring the unknown in space and the atmosphere is at the heart of who we are at NASA," stated Thomas Zurbuchen, associate administrator of the Science Mission Directorate at NASA headquarters in Washington. "Understanding the data we have surrounding unidentified anomalous phenomena is critical to helping us draw scientific conclusions about what is happening in our skies. Data is the language of scientists and makes the unexplainable, explainable."

The agency and most sensible people readily admit the existence of unidentified phenomena in the skies. After all, such phenomena—flying Tic Tac–shaped apparitions off the coast of California, for example, or an errant "cruise missile" over Las Cruces—have been captured on audio and video. (Grainy video, unfortunately, but video nonetheless.) But few are willing to leap from this proposition to the notion that the objects are spaceships operated by extraterrestrials or, possibly, operated by artificial intelligence *created* by extraterrestrials. Still, the notion is hard to resist, containing as it does so many intriguing possibilities for doom, dread, and existential unrest. A final report issued by NASA's sixteen-member panel in September of 2023 declined to find any extraterrestrial explanation for certain reported phenomena but admitted that some data are at present unexplainable. The result wasn't particularly surprising, but it did manage to upset some of the more partisan UFOlogists who had been following both the proceedings of the panel and a previous summer's worth of incendiary testimony before Congress regarding an alleged DOD alien-recovery operation.

As Charles Cockell puts it in his 2022 book *Taxi from Another Planet*, "Science and science fiction have always danced around each other as if in a waltz, and never so much as in the area of alien life." The waltz will no doubt continue as the universe continues to supply new bases for speculation. Here's a recent example. In 2017 a cigar-shaped reddish object some five hundred meters long entered the ken of astronomers and seemingly *accelerated* as it passed through our solar system. This change in velocity was a phenomenon difficult to square with our understanding of physics. The blue whale–shaped rock was named 'Oumuamua, a Hawaiian word meaning "scout," or "messenger from afar arriving first." It is believed to be the first of its kind ever recorded—not a planet, moon, asteroid, or comet but rather a new variety of "interstellar traveler," origin unknown.

Intrigued by the object's anomalous shape and incongruous acceleration, Harvard astronomer Avi Loeb theorized that 'Oumuamua might be an extraterrestrial artifact. He suggested that it could be a "starship" of sorts, manufactured long ago and far away. The supposition has found little support in the scientific community. Loeb has acquired a reputation as a sort of Captain Ahab of the alien visitation community. He might as well have claimed to see Santa Claus in the skies over Cambridge. Nevertheless, the Scout is a helpful reminder that the cosmos is going to continue to surprise us. And Loeb, for one, is undaunted by the doubters. He's convinced that our first contact with aliens will look less like the movie *War of the Worlds* and more like an episode of *Antiques Road Show*, with experts poring over obscure evidence of off-world machinery. In 2021 he founded the Galileo Project, an organization dedicated to searching for technological artifacts of extraterrestrial visitation. So far, no dice. But the effort continues.

The truth is out there, people. And it's likely to be strange.

Eleven Fearless Space Predictions

1. Humankind will return to the moon in 2028, though it's unclear whose rocket will take us there. It might be Chinese.

2. The first asteroid will be mined in 2032, with marketable quantities of tungsten, nickel, and gold returned to Earth in 2033. Get ready for asteroid earrings!

3. The first nuclear fission–powered spaceship will fly in 2035. Submarines have been using nuclear propulsion for decades. Why can't crewed spaceships?

4. Spurred by low-Earth orbit data obtained by NASA and ESA satellites, including imaging of major methane gas leaks, humanity will in 2037 finally make significant progress in reducing man-made greenhouse gas emissions.

5. Human beings will walk on Mars in 2039. It's unclear what language they will speak. See Fearless Space Prediction No. 1, above.

6. Men and women will begin living on Mars in 2044. Potatoes will probably be involved.

7. Settlements on Mars will eventually be populated mostly by robots, which can function—indeed, *thrive*—in the high-radiation, low-oxygen atmosphere of the Red Planet.

8. Working together in early 2045, scientists and military personnel will deflect a large asteroid from its collision course with Earth, thus preventing a cross-continental environmental disaster—and millions of human deaths.

9. In 2047 researchers will finally manage to create a nuclear fusion engine, opening up the possibility of human travel beyond Jupiter.

10. In 2050 life will be discovered in the subsurface oceans of Saturn's moon Enceladus. Biologists will struggle to classify it. They will be happy anyway.

11. By the end of the century, humanity will contact—or be contacted by—extraterrestrial intelligence. Let's try to get our affairs in order, shall we?

20

Why It's Worth It

Onward.

Occasionally, someone tries to resurrect the old either/or. *Why space?* they ask. Why are American taxpayers asked to foot the bill for space exploration— some $25 billion for the 2024 fiscal year, less than one half of 1 percent of the federal budget—when we could be making things better here on our planet? After all, the argument goes, with enough money, we could end poverty, eradicate the causes of world hunger, heal the injured earth. Why set our sights on shadowy gems in the outer dark when our battered blue jewel so desperately needs our help?

There was a time when the question made sense.

Sixty years ago America was locked in a dangerous and expensive battle for geopolitical supremacy with the Soviet Union. Its most visible competition was a race to "conquer" the barren surface of the moon, a do-or-die technological slugfest that seemed to return about as many material benefits as conquering the Gobi Desert. Getting there was an astonishing engineering accomplishment. It paid dividends in international prestige that the United States is still collecting today. Nevertheless, the Apollo years were a parade of horribles. Martin Luther King Jr. and Bobby Kennedy were assassinated. We fought a vicious and possibly unwinnable war in Southeast Asia. The bald eagle teetered on the edge of extinction, and Cleveland's Cuyahoga River caught fire on multiple occasions. And though we made it to the moon, we weren't really clear on what should come next. It took us years to figure out that the moon has resources—*water*, for example—that could actually be harvested.

Things have changed. The United States still suffers from simmering racial resentments, to be sure. It takes time to suture up centuries of inequality. But the notion of space exploration as the pet project of white privilege makes less sense in an age when former president Barack Obama, an avowed *Star Trek* fan, advocates for missions to asteroids, and his handpicked NASA administrator, Charlie Bolden, was just one of many people of color to have ventured into the cosmos on the space shuttle or a Russian rocket.

And yes, the world is in big environmental trouble—bigger in fact than when we faced localized problems like smoldering rivers and LA smog. But spending money on space isn't making that problem worse. Space technology isn't detracting from conservation efforts. It's assisting them. In fact, data and imagery from space have been huge factors in generating assessments of terrestrial environmental problems as well as interest in solving them. For example, NASA researchers were at the forefront of the data collection efforts in the eighties that showed how chlorofluorocarbons were destroying Earth's ozone layer, an atmospheric blanket that protects the planet from harmful ultraviolet radiation. International efforts to restrict use of such compounds has resulted in a remarkable restoration of the ozone barrier. NASA's James Hansen was among the first scientists to sound the alarm on climate change, providing vivid testimony on the topic to Congress in 1988.

The agency currently operates a broad climate research program. Among the many areas NASA studies are solar activity, rising sea levels, temperatures in the atmosphere and ocean, the health of the ozone layer, air pollution, and changes in land and sea ice. Asteroid defense, an endeavor initiated with the DART mission, is perhaps the greatest environmental protection effort ever undertaken, since a large asteroid's impact with Earth could wipe out every elephant, tiger, and Instagram influencer on the planet. Even during the frequently anti-science days of certain recent administrations, NASA maintained its website warning of dangers due to our overuse of fossil fuels. The agency continues to ring the alarm bell on climate change. Whether enough of us will act on such warnings is another issue altogether.

Most importantly, though, the reason why the whole space vs. social spending issue no longer makes sense is that space is no longer a hypothetical. We're no longer trying to get to space. We're *in* it. It's happening now, not at some point in the future. America's satellite-based Global Positioning System, GPS, is perhaps the most important byproduct of our early years of space exploration. It provides hundreds of millions of people—tourists and truck drivers, cargo plane pilots and cruise ship captains—with navigational data on a daily basis. Communication satellites enable us to talk to each other and exchange information wherever we are in the world. NOAA weather satellites warn us of storm fronts, heat waves, hurricanes, and other hazards. Observation platforms operated by environmental groups monitor the emission of methane and other greenhouse gases. Private companies like Blue Origin and Axiom are offering space tourism flights similar in spirit, if not in expense, to the rides that biplane-driving barnstormers sold to eager farm boys a hundred years ago. And if eighty-one-year-old Wally Funk can enjoy a jaunt above the Kármán line, why can't we? Space is *here*. We're in it. The question is, what do we do with it?

America as Role Model

The importance of continuing, and indeed substantially increasing, America's spending on space exploration is clear. There are several reasons. First, we need to preserve our international lead in promoting the understanding of Earth, the solar system, and the universe. The prestige associated with innovations like GPS, accessible to everyone, and the use of the James Webb and Hubble Space Telescopes, available to researchers all over the world, are invaluable. Despite our internal political squabbles about climate change and vaccinations, gun violence and reproductive rights, the United States is still seen as a nation of science and technology pioneers—Thomas Edison and Jonas Salk, Elon Musk and Gladys West.

The United States is certainly not the only country that has sent people or probes into space. It's a rapidly expanding club that now includes Russia, China, India, Japan, Israel, the United Arab Emirates, and the many nations of the European Space Alliance. But we are, still and perhaps increasingly, the nation that seems most interested in searching the cosmos because we simply want to know what's out there, the nation that believes at some level that the quest to understand the universe is integral to human consciousness.

It's an expensive effort, to be sure. But the rewards are tremendous. Great nations do great things. The huge public engineering project that was Apollo brought some four hundred thousand Americans—California Democrats and Alabama Republicans, secret hippies and ex-Marines, male and female, old and young—together to work toward the same goal. It generated a sort of social and cultural cross-pollination that helped to define what the country is and what it's capable of. The continuing chance to work in America's open, ambitious, and merit-based space program is a powerful magnet for international talent, attracting bright young people like India's Kalpana Chawla and Costa Rica's Franklin Chang-Díaz from all over the globe to study at American universities and explore employment possibilities on the high frontier. To squander this leadership position would be a tragic waste of international goodwill. Influence can be exerted through the barrel of a gun. Inspiration arises more subtly—and sometimes through the lens of a telescope. There has to be some unspoken moral suasion in the fact that in 2022, while Russia was killing civilians and destroying cities in neighboring Ukraine, NASA was figuring out how to deflect asteroids from a collision course with Earth, an enterprise that we may all be thankful for one day.

Space as the New Klondike

Space is increasingly understood as a huge, practically infinite set of resources to be harvested, from solar energy in low-Earth orbit to helium-3 on the moon to gold, nickel, and lithium on any number of asteroids in our solar system. The United States and Japan have already started sampling asteroids. Mining operations on the moon, Mars, and various smaller solar system objects are not a matter of if, but when. Losing out on the acquisition of these resources would be a big mistake. And losing out is a possibility. Not only will mining in space be technologically challenging. There will also be competition.

From difficult beginnings in the fifties and sixties, China has emerged as a power in space second only to the United States—and it's closing fast. China has built its own space stations, probed the far side of the moon, and announced plans for crewed lunar and Mars missions. Japan, meanwhile, has visited two asteroids. India, whose *Chandrayaan-1* probe discovered water ice on the moon in 2009, has landed near the moon's south pole and plans to send astronauts to our lunar satellite soon. Russia, Canada, and the European Space Agency are capable players in the accelerating space race, limited more by a lack of funds than any want of expertise. And finally, private interests will eventually figure out ways to acquire and profit from resources present on asteroids and other planets and moons. If the United States wants to retain any measure of influence in such efforts, we need to be a spacefaring nation— and ideally, the *preeminent* spacefaring nation, as Great Britain was the preeminent seafaring power of the nineteenth century. Not to colonize, of course. But to compete. It turns out the space race isn't over. In fact, it's just begun.

Spiritual Benefits?

One of the less tangible benefits of space travel is its effect on our collective psyche—and possibly even our souls. First coined by self-styled "space philosopher" Frank White in 1985, the term *overview effect* refers to a broadening of appreciation for Earth and its inhabitants that many—perhaps most—astronauts experience during spaceflight. The effect derives from seeing the planet as a whole rather than as simply a set of immediate surroundings. It's unclear whether this is a unique type of "cognitive shift" or just the sort of practical appreciation of our global oneness that one can get also from traveling terrestrially. Astronaut Bruce McCandless II returned with strong symptoms of the effect, which in his case manifested as a shift in perspective toward globalism as a result of viewing the planet from space. As he later told *National Geographic*, "As a blanket statement, I believe I'm OK in saying that just about everyone who has flown in space and looked down on the Earth has altered their perception of it. And the prevailing feeling seems to be that when we look down from space, we really can't see the political subdivisions, and we wonder, why [we]—meaning everybody on spaceship Earth—why we can't learn to work with each other and get along."

Other space travelers have been explicitly religious in describing their feelings. In a lecture at Columbia University, Mike Massimino described his first space walk with awe. As he looked down on Earth, he thought, "This is a secret; this is too beautiful for humans to see." His initial instinct was to turn his head away from the view. "This must be the view from heaven," he mused. "This is what God sees." Apollo astronaut Jim Irwin described having a religious experience on the moon. In a 1991 article published after Irwin's death, the *New York Times* reported that he would often tell church groups he "felt the power of God as I'd never felt it before" in that moment.

Others describe the change in perspective they experienced in space in less easily definable terms. On *Apollo 14*, astronaut Ed Mitchell and his colleague Alan Shepard traversed the Fra Mauro region of the moon and trekked toward Cone Crater to gather geological samples that they hoped might reveal information about the moon's inner structure. According to a biographical sketch published by the *New York Times* after Mitchell's death in 2016, as the mission's command module traveled homeward, Mitchell watched the earth, moon, and sun passing by the window of the slowly rotating spaceship. Looking out into space, Mitchell later recalled, "I realized that the molecules of my body and the molecules of the spacecraft had been manufactured in an ancient generation of stars." Nothing in his military or NASA training had equipped him for a sudden discovery of the oneness of all things. "It was a subjective visceral experience accompanied by ecstasy," he would later explain. After returning to Earth, he left NASA and founded the Institute of Noetic Sciences, which advocates exploring the universe by means of inquiry that lay outside of science and religion. Mitchell, says the *New York Times*, "sought out South American shamans and Haitian Vodou priests, promoted the benefits of Tibetan Buddhist lucid dreaming, and visited the homes of people who claimed their children could bend spoons with their minds." Whether he located any actual spoon benders is unclear, but the influence of space travel on his thinking is not.

Space changed him.

Admittedly, other astronauts deny having strong religious or spiritual feelings at all while traveling in space. But evidence does exist that the act of viewing Earth from out there sparks feelings of commonality in most and moments of ecstasy in some. Whether this is enough to justify the enormous costs and risks involved in space travel is, of course, another question. But it's a question that we all ought to have the benefit of asking for ourselves.

Planet B

Some people imagine that we will journey to distant galaxies because we're composed of cosmic particles—hydrogen, carbon, and other materials created by ancient stellar infernos. We are, they say, star dust, and our yearning to sail the heavens is at root an elemental nostalgia for *home*. According to this view, venturing into outer space is predestined, our answer to a call that's older than our bones.

We don't buy it. A little of that sort of thing goes a long way in our book, and since this *is* our book, we'll stop it here. We're also made of water, but not everyone likes to go to the beach. And besides, if we're made of "star stuff," as Carl Sagan called it, so is everything around us. Trees. Apples. Ryan Gosling. *Why leave?* We humans have an affinity for sun and mud and the feel of grass beneath our feet. We like to lie in hammocks and pretend we're pirates. We're not constructed for constant cold and infinite dark and the sterile white of artificial light. Nevertheless, some of us will leave the planet anyway, not because of some mitochondrial itch but because we are pulled by the lure of potential riches, propelled by national rivalries or venturing outward in search of another oasis, an extra Earth, and the glory of its discovery. We'll go because we're human. Because our brains won't let us rest. And because deep down, we know paradise has an expiration date. Our Earth won't last forever.

And this is perhaps the most compelling reason for learning to live and work outside our atmosphere. We need space in order to protect the earth and our favorite inhabitants: *us*. The way futurist Gerard K. O'Neill saw it, a move into space colonies would help Earth's environment by easing the planet's population load and slowing humanity's quest for natural resources, many of which could be found and refined in space. Other advocates say we need an alternate Earth, a plan B—not as an escape hatch, as author Charles Cockell points out, but as an insurance policy. We would be wise to plan, for example, for the appearance of an asteroid as large as the one that pierced the planet sixty-six million years ago, wiping out our world's reptilian overlords and a substantial portion of the rest of Earth's flora and fauna as well.

The odds of such an event happening in the near future are small, but they're worth taking seriously. Though a planet-killing asteroid is mostly the stuff of nightmares, less destructive meteor impacts occur on a fairly regular basis. We need to know how to detect incoming asteroids as early as possible and how to deflect them once we know they're a menace. The United States and China are both working on projects with these goals in mind. The agent of our destruction might not be an asteroid. It might be a comet. It might be a volcanic eruption. It could even be a pandemic, a future Black Death that sneaks past our beleaguered antibiotic defenses and destroys humanity's ability to care for itself.

If worse comes to worst and Earth is no longer a viable habitat, we need to have a cabin in the woods for the preservation of humanity itself—whether it's a single crewed base on Mars or a robust system of off-world space stations, lunar installations, and mining operations all through the solar system. Anyone who's ever seen Captain James T. Kirk wrestle a Gorn knows there are planets out there suitable for human habitation, grouchy space lizards notwithstanding. In our collective imagination, we've seen humans settle asteroids, colonize Mars and Jupiter, thrive on planets orbiting distant stars. But where, exactly, might these planets be? The solar system offers limited options. While it appears possible to plant colonies on Mars, it won't be easy. Mars isn't exactly welcoming. Most of it is frigid. Wind-blown dust gets into everything. There's essentially no oxygen in the atmosphere, which is dominated by carbon dioxide. In fact, there's very little atmosphere, *period*, so there's not much of a screen against solar radiation—a big problem for us thin-skinned humans. On Mars we'd have to find or, more likely, make our own water and use it not only for drinking but also for growing any sort of food we want to produce. The moon has all these problems, to a greater degree. To paraphrase Gertrude Stein's remark about Oakland, there's no *there* there. Venus? Better to live in a catalytic converter. Jupiter? Every day would feel like a car wreck. Saturn's moons? Bring a hoodie; Titan has a surface temperature of −292 degrees Fahrenheit. Not even the Packers could play there.

So if we're looking for a possible future home—a *pleasant* future home, at any rate—we need to search farther afield. We now know of numerous planets outside of our solar system. The count of such spheres exceeded five thousand in early 2022, when sixty-five new worlds were added to NASA's Exoplanet Archive. The archive, a sort of library of exotic orbs, has stringent inclusion requirements, which means that the thing you saw in the sky while you were camping in Arkansas last October probably isn't going to qualify. As of August 2024 the tally stands at 5,747, with thousands more "candidate" planets waiting in the wings.

Given that theorists have calculated that we have some hundred thousand million stars in our galaxy, which is only one galaxy out of possibly *trillions* in the universe, scientists and their scruffier cousins, science fiction writers, have long assumed that other planets existed. It's just that the technology we needed for finding them is of recent vintage. In fact, every one of the thousands of worlds we know about has been discovered since 1992, when the first exoplanets (Poltergeist and Phobetor) were spotted orbiting a pulsar, designated as PSR B1257+12, in the constellation Virgo. The planet 51 Pegasi b, the first planet orbiting a "main sequence" (i.e., hydrogen-burning) star, was spotted in 1995. The closest exoplanet we know of is Proxima Centauri b, at around 4.25 light years from Earth. Among the farthest is OGLE-2014-BLG-0124Lb, at around *thirteen thousand* light years from our world. And while we can certainly continue to ogle OGLE, we won't be visiting any time soon.

It's difficult to generalize about these worlds, though of course that's exactly what scientists like to do. They paint odd pictures. For example, massive "hot Jupiters" orbit so close to their sun that they are deformed by its gravitational pull. One, Wasp 121b, is thought to be home to atmospheric metals that grow so hot due to this solar proximity that they vaporize and form clouds of metallic elements. When they reach the boundaries of the "cool" side of the planet, these elements turn to liquid, which means that the planet could see rains of liquid ruby and sapphire. Some planets are so large and contain so much carbon that their hyper-pressurized cores could be made of diamond. One planet is thought to have so much silicate in its scalding atmosphere that its skies rain molten glass. Another, a so-called hot Neptune, rains titanium. Other exoplanets have seas of water vapor so hot that they could melt your spacesuit. A relatively small number have wandered away from their stars and are on long, lonely journeys to nowhere.

Finding planets is exciting. Even more tantalizing is finding planets that look like our own—orbs that might be capable not only of allowing for human settlement but of supporting it. What characteristics would such a planet have? First, it would need to be around the same size as our home globe so that it would have a similar gravitational feel. It would also need to be in its star's "habitable zone"—at a distance from the star where the planet would be warm enough to support life, but not *too* warm. This means that the planet would need to either orbit a star similar in size to ours or be closer to a smaller star or farther away from a larger one. An atmosphere would be important to filter out solar radiation, and an atmosphere with air would be ideal. Finally, water in an easily usable form—that is, a surface liquid—would be helpful, though if it weren't present on the surface, water could be obtained through drilling or ice mining. A planet that fits all these criteria can be described as a "Goldilocks" planet, neither too large nor too small, too cold nor too hot, too wet nor too dry.

There are lots of other variables, of course, some of which we can't imagine yet, because we haven't been there. For example, are there other life forms on the planet, and how would they tolerate coexistence with human beings? This is the stuff of Asimov, Heinlein, and an acid-filled whatsit scuttling through the air ducts of the *Nostromo* in the sci-fi classic *Alien*. It brings us back to Kirk, Spock, and the gang. But the fact remains that given the large numbers involved, there are—there pretty much *have* to be—Goldilocks planets out there. Some of them might be suitable locations for a New New Mexico, or a New New York. No one wants to imagine a New New Jersey, but *it's* out there too. Perhaps someday we'll see it.

Sitting on a Rooftop

Space is a stunningly hostile environment. It freezes and burns and bombards the body with tiny radiation grenades. Prolonged travel in space sucks the marrow from the human skeleton. Our eyeballs flatten. Our muscles deteriorate. We gradually lose the ability to stand on our own two feet. Furthermore, though the search is young, we know of no planet, moon, or asteroid besides Earth that supports any type of life. Astrobiology is a fascinating topic, but so far it's a study without a subject, an exercise in speculation.

Wherever we go in our solar system, we're going to have to create our own environment. In doing so, we'll need to contend not only with the usual human needs—oxygen, water, food, and decent Wi-Fi—but also with the constant threat of cosmic radiation on any celestial body that lacks the protective atmosphere possessed by Earth. And that's just once we get there. The distances involved in traveling to celestial destinations are staggering. With current technology, even a voyage to Mars will take months, in close quarters, with limited protection from radiation exposure. The next logical step, a trip to Jupiter's apparently water-rich moons, will take *years*, with little chance for support or rescue along the way and even less once there. It's unclear how fit the archetypal astronaut would be for such a journey. Some scientists say women, just as intelligent as men but generally smaller and lighter, might make a better crew. Others wonder why we would send humans at all when robots are close to being able to do everything a person can do, all without eating, sleeping, emitting noxious waste materials, or arguing about college football. We can explore with space telescopes. We can discover new worlds with automated probes. So it's high time to stop thinking of space crews as Neil, Buzz, and Mike. The first band of plucky explorers to make it to another planet might well consist of Neil, Siobhan, fourteen R2D2s, a knowledge-spewing but occasionally wildly inappropriate AI program, and a Boston Dynamics dog-bot named Bodhi. Our first settlements might be underground, protected from radiation, where our astronauts (and astro-bots) would spend their time harvesting water, oxygen, and rocket propellant, growing food, and mining for precious metals.

Humanity's relationship with the cosmos could end up like our interactions with Antarctica. Science teams might someday spend extended periods of time on Mars, surrounded by dust, just as they currently do at McMurdo

Station, surrounded by ice. Researchers would do their work and go home. Tourists would visit and quickly leave. We could maintain a presence on the Red Planet without thinking much about it.

Or maybe our presence in space would be like our presence on the ocean. It's an obvious comparison, especially now that scientists have confirmed the presence of gravitational "waves" rippling across the universe like shudders in a silent sea. Science fiction hasn't always looked up at the stars. As far back as Jules Verne, it has also peered into the depths. No one is clamoring for a new Atlantis these days, and no one lives in an underwater city, though the residents of New Orleans are getting close. There are subsurface hotels and restaurants, to be sure, but these are tourist attractions rather than productive habitats. Nevertheless, we traverse the oceans all the time. We take food and energy from the seas in ways and at scales that would have been unimaginable five hundred years ago—and which may well prove to be unsustainable in the very near future. Men and women go down to the sea in ships. They work for a period in or on this dangerous environment, and then they return. So will the cosmos be like the seas, traveled but not inhabited by humanity? Or will we find some way to make it a home, at least for a portion of our species? Will we find evidence of life on some distant planet? And then, eventually, meet up with it? And when will we claim our rightful place in the great parliament of sentient species currently being called to order somewhere in the Vega constellation?

We may find out sooner than we think. The question we face as a nation is no longer Earth vs. space. Low-Earth orbit, impossible only a hundred years ago, already seems more like part of the planet than it does of the great expanse beyond. The rest of the solar system is getting closer, year by year. Given the ambitions of so many other countries and companies around the globe, our choice is clear. Either America retains its lead in exploring the heavens or other nations will seize the torch. So perhaps it should finally be revealed that this little treatise is not just an introduction to a challenging, infinitely interesting frontier, to worlds without end and a universe varied and spectacular beyond all understanding. It's also an argument in favor of remembering Explorer 1 and the *Voyager* probes, of taking up the mantle of Robert Goddard and Max Faget, Eileen Collins and Ron McNair, and pushing out into the cosmos.

And finally, it's an invitation to you, our reader, to get enthused. To get involved. Join a space society. Watch a launch—or a landing. Write to your congressperson in support of NASA's anti-asteroid efforts. Heck, apply to become an astronaut if you feel the inner fire. Anyone who's ever sat on a rooftop on a frigid night and watched the stars drift overhead like advertisements for adventure knows that we're bound for forever, for discoveries and disappointments and a destiny that we can't even imagine at the moment. That's where we've always been bound. But someone's going to have to lead us there.

And maybe that someone is you.

Sources

Books

Adams, Douglas. *The Hitchhiker's Guide to the Galaxy.* New York: DelRay, 1995.

Allen, Joseph P. *Entering Space: An Astronaut's Odyssey.* New York: Stewart, Tabori & Chang, 1984.

Bagby, Meredith. *The New Guys: The Historic Class of Astronauts that Broke Barriers and Changed the Face of Space Travel.* New York: William Morrow, 2023.

Berkowitz, Bruce, and Michael Suk. *The National Reconnaissance Office at 50 Years: A Brief History.* Washington DC: The National Reconnaissance Office, 2019.

Bova, Ben. *The High Road.* Boston: Houghton Mifflin, 1981.

Bowen, Bleddyn E. *Original Sin: Power, Technology, and War in Outer Space.* Oxford: Oxford University Press, 2023.

Burgess, Colin. *Soviets in Space: Russia's Cosmonauts and the Space Frontier.* London: Reaktion, 2022.

Burrough, Brian. *Dragonfly: NASA and the Crisis aboard Mir.* New York: Harper Collins, 1998.

Burrows, William E. *This New Ocean.* New York: Random House, 1998.

Cernan, Eugene, and Don Davis. *The Last Man on the Moon: Astronaut Eugene Cernan and America's Race in Space.* New York: St. Martin's, 1999.

Chaikin, Andrew. *A Man on the Moon: The Voyages of the Apollo Astronauts.* New York: Penguin, 1994.

Chudwin, David. *I Was a Teenage Space Reporter.* New York: LID, 2019.

Clary, David. *Rocket Man: Robert H. Goddard and the Birth of the Space Age.* New York: Hachette, 2004.

Cockell, Charles E. *Taxi from Another Planet: Conversations with Drivers about Life in the Universe.* Cambridge MA: Harvard University Press, 2022.

Collins, Eileen M., and Jonathan H. Ward. *Through the Glass Ceiling to the Stars: The Story of the First Woman to Command a Space Mission.* New York: Arcade, 2022.

Collins, Michael. *Carrying the Fire: An Astronaut's Journeys.* New York: Farrar, Straus & Giroux, 2019.

Cunningham, Walter. *The All-American Boys: An Insider's Look at the U.S. Space Program*. New York: ipicturebooks, 2009.

Frank, Adam. *The Little Book of Aliens*. New York: Harper 2023.

Gallentine, Jay. *Infinity Beckoned: Adventuring Through the Inner Solar System, 1969–1989*. Lincoln: University of Nebraska Press, 2016.

Garver, Lori. *Escaping Gravity: My Quest to Transform NASA and Launch a New Space Age*. New York: Diversion, 2022.

Gilbert, Martin. *The Second World War*. New York: Henry Holt, 1989.

Grush, Loren. *The Six: The Untold Story of America's First Women Astronauts*. New York: Simon & Schuster, 2023.

Hansen, James R. *First Man: The Life of Neil A. Armstrong*. New York: Simon & Schuster, 2012.

Harrison, Jean-Pierre. *The Edge of Time: The Authoritative Biography of Kalpana Chawla*. Littleton NH: Harrison, 2011.

Hersch, Matthew. *The Making of the American Astronaut*. New York: Palgrave MacMillan, 2012.

Launius, Roger D. *NASA: A History of the U.S. Civil Space Program*. Updated ed. Malabar FL: Krieger, 2001.

Liu, Cixin. *The Three-Body Problem*. New York: Tor, 2016.

Loeb, Avi. *Extraterrestrial: The First Sign of Intelligent Life beyond Earth*. New York: Mariner, 2021.

Logsdon, John, ed. *The Penguin Book of Outer Space Exploration: NASA and the Incredible Story of Human Spaceflight*. New York: Penguin, 2018.

Lucid, Shannon. *Tumbleweed: Six Months Living on Mir*. MkEk, 2020.

Mailer, Norman. *Of a Fire on the Moon*. New York: Random House, 2014.

McCandless, Bruce, III. *Wonders All Around*. Austin TX: Greenleaf, 2021.

Muir-Harmony, Teasel. *Operation Moonglow: A Political History of Project Apollo*. New York: Basic Books, 2020.

Mullane, Mike. *Riding Rockets: The Outrageous Tales of a Space Shuttle Astronaut*. New York: Scribner, 2007.

Neufeld, Michael J. *Spaceflight: A Concise History*. Cambridge MA: MIT Press, 2018.

———. *Von Braun: Dreamer of Space, Engineer of War*. New York: Vintage, 2007.

O'Leary, Brian. *The Making of an Ex-Astronaut*. Boston: Houghton Mifflin, 1970.

O'Neill, Gerard K. *The High Frontier: Human Colonies in Space*. New York: William Morrow, 1976.

Parsons, Paul. *A Hopeful Vision of the Human Future*. New York: Simon and Schuster, 1981.

———. *Space Travel: Ten Short Lessons.* Baltimore MD: Johns Hopkins University Press, 2020.

Pendle, George. *Strange Angel: The Otherwordly Life of Rocket Scientist John Whiteside Parsons.* New York: Mariner, 2006.

Pyle, Rod. *Amazing Stories of the Space Age: True Tales of Nazis in Orbit, Soldiers on the Moon, Orphaned Martian Robots, and Other Fascinating Accounts from the Annals of Spaceflight.* Dallas TX: Prometheus, 2017.

Pyne, Steven J. *Voyager: Seeking Newer Worlds in the Third Great Age of Discovery.* New York: Viking, 2010.

Roach, Mary. *Packing for Mars: The Curious Science of Life in the Void.* New York: W. W. Norton, 2022.

Slayton, Donald K., and Michael Cassutt. *Deke! U.S. Manned Space from Mercury to the Shuttle.* New York: Forge, 1994.

Sullivan, Kathryn D. *Handprints on Hubble: An Astronaut's Story of Invention.* Cambridge MA: MIT Press, 2019.

Teitel, Amy Shira. *Breaking the Chains of Gravity: The Story of Spaceflight Before NASA.* New York: Bloomsbury Sigma, 2018.

Tribbe, Matthew D. *No Requiem for the Space Age: The Apollo Moon Landings and American Culture.* New York: Oxford University Press, 2014.

Walker, Stephen. *Beyond: The Astonishing Story of the First Human to Leave Our Planet and Journey into Space.* New York: Harper, 2021.

Weir, Andy. *Project Hail Mary.* New York: Ballentine, 2021.

Weitekamp, Margaret A. *Space Craze: America's Enduring Fascination with Real and Imagined Spaceflight.* Washington DC: Smithsonian Books, 2002.

White, Rowland. *Into the Black.* New York: Touchstone, 2016.

Wienersmith, Kelly, and Zach Wienersmith. *A City on Mars: Can We Settle Space, Should We Settle Space, and Have We Really Thought This Through?* New York: Penguin, 2023.

Wilgus, Alison, and Wyeth Yates. *The Mars Challenge: The Past, Present, and Future of Human Spaceflight.* New York: First Second, 2020.

Wolfe, Tom. *The Right Stuff.* New York: Farrar, Straus and Giroux, 1979.

Worden, Al, and Francis French. *Falling to Earth: An Apollo 15 Astronaut's Journey to the Moon.* Washington DC: Smithsonian Books, 2012.

Periodicals and Online Articles

Aviation Week & Space Technology. "Martin Converts USAF Titan 2 to Launch Vehicle for Placing Defense Payloads into Polar Orbit." 10 August 1987.

254 | SOURCES

———. "U.S. Action on Commercial Space Policy Criticized by Current, Former Administration Officials." 29 September 1986.

Baltimore Sun. "Order Lets Tsien 'Self-Deport' to China." 13 September 1955.

Barlow, Nathan. "Thomas Merton Lecture: Views from Space." *Christian Union: The Magazine,* Spring 2019. www.christianunion.org/publications-media/christian -union-the-magazine/past-issues/spring-2019/2326-thomas-merton-lecture- %E2%80%9Cviews-from-space%E2%80%9D.

Bartels, Meghan. "The Kardashev Scale: Classifying Alien Civilizations." *Science,* 10 December 2021. www.space.com/kardashev-scale.

Bedford (IN) Daily Times-Mail. "Gus Grissom's Wife Files $10 Million Suit." 19 January 1971.

Benedict, Howard. "Shuttle Chief Accelerating NASA's Slow, Conservative Approach." *Mobile Press,* 25 January 1983.

Berger, Eric. "Heinlein and Clark Discuss the Moon Landings as They Happen." *Ars Technica,* 21 December 2016. https://arstechnica.com/science/2016/12/heinlein -and-clarke-discuss-the-moon-landings-as-they-happen/.

Billings, Lee. "Astronomer Avi Loeb Says Aliens Have Visited, and He's Not Kidding." *Scientific American,* 1 February 2021. www.scientificamerican.com/article /astronomer-avi-loeb-says-aliens-have-visited-and-hes-not-kidding1/.

Calgary Herald. "Soviets Unaware That U.S. Satellite Was Spying." 23 November 1978.

Chang, Kenneth. "Hubble Space Telescope Spots Earliest and Farthest Star Known." *New York Times,* 30 March 2022. www.nytimes.com/2022/03/30/science/hubble -star-big-bang.html?action=click&pgtype=Article&state=default&module =styln-hubble-telescope&variant=show&region=MAIN_CONTENT_1& block=storyline_top_links_recirc.

Chicago Tribune. "New Era Ushered in by Shuttle." 15 April 1981.

Clegg, Brian. "What Is Gravity? A Guide to Nature's Most Mysterious Force (And What We Still Don't Know)." BBC Science Focus, 21 November 2021. www .sciencefocus.com/space/gravity/.

Cook, William J. "Straighten Up and Fly Right." *U.S. News & World Report,* 16 July 1990.

Cooper, Henry S. F. "Letter from the Space Center." *New Yorker,* 18 February 1985.

Daily Arkansas Gazette. "Experts Dumbfounded." 24 March 1918.

David, Leonard. "Anatomy of a Spy Satellite." Space.com, 3 January 2005. www.space .com/637-anatomy-spy-satellite.html.

Davidson, Carl. "Andover Slept till Telstar Arrived." *The Times* (Munster IN), 2 September 1962.

SOURCES | 255

Drake, Nadia. "First Person to Walk Untethered in Space Gives a Final Interview." *National Geographic*, 7 February 2018. www.nationalgeographic.com /news/2018/02/first-untethered-spacewalk-bruce-mccandless-astronaut-space -science/.

Easterbrook, Gregg. "Big Dumb Rockets." *Newsweek*, 17 August 1987.

———. "NASA: What Goes Up." *Los Angeles Times*, 22 July 1990.

Erwin, Sandra. "Military Surveillance Constellation Fuels Debate over Who Calls the Shots." SpaceNews.com, 8 May 2024. www.spacenews.com/military-surveillance -constellation-fuels-debate-over-who-calls-the-shots/.

———. "U.S. Space Force Wary of China's Expanding Spy Satellite Fleet." SpaceNews .com, 30 January 2024. www.spacenews.com/u-s-space-force-wary-of-chinas -expanding-spy-satellite-fleet/.

Evans, Ben. "The Saddest Moment: The Story of America's First Spacewalk." AmericaSpace.com, 31 May 2015. www.americaspace.com/2015/05/31/the -saddest-moment-the-story-of-americas-first-spacewalk-part-2/.

Fisher, Kristin. "Inside the Secretive Process to Select the First Astronauts for NASA's Next Moon Mission." CNN.com, 29 January 2023. www.cnn.com/2023/01/29 /world/nasa-artemis-moon-secretive-crew-selection-process/index.html.

Fort Lauderdale (FL) News. "Planned Space Shuttle Transport Called Revolutionary." 12 May 1971.

Fox, Margalit. "Max Faget, 83, Pioneering Aerospace Engineer, Dies." *New York Times*, 12 October 2004. www.nytimes.com/2004/10/12/obituaries/maxime -faget-83-pioneering-aerospace-engineer-dies.html.

Franklin, Jon. "Heavens to Put on Show of a Lifetime." *Baltimore Sun*, 8 July 1973.

Gamillo, Elizabeth. "Hubble Space Telescope Spots Largest Comet Ever Discovered." *Smithsonian Magazine*, 15 April 2022. www.smithsonianmag.com/smart-news /hubble-space-telescope-spots-largest-comet-nucleus-ever-discovered-180979924/.

Garofolo, Meredith. "Lost Satellite Found after Orbiting Undetected for 25 Years." Space.com, 6 May 2024. www.space.com/lost-satellite-found-us-space-force -data.

Gellman, Barton, and Greg Miller. "Black Budget Summary Details U.S. Spy Network's Successes, Failures, and Objectives." *Washington Post*, 29 August 2013.

Goldstein, Phil. "How the IBM 7094 Gave NASA and the Air Force Computing Superiority in the 1960s." *FedTech Magazine*, 11 October 2016. https://fedtechmagazine .com/article/2016/10/how-ibm-7094-gave-nasa-and-air-force-computing -superiority-1960s#:~:text=The%207094%20was%20first%20introduced,the %20space%20program%20from%20NASA.

SOURCES

Gray, Richard. "What Does Spending More than a Year in Space Do to the Human Body?" BBCFuture.com, 27 September 2023. www.bbc.com/future/article/20230927-what-a-long-term-mission-in-space-does-to-the-human-body.

Greenville News. "Skylab Crew Takes Over Space Endurance Record." 19 June 1973.

Hall, Loretta. "Setting the Record: Fourteen Months aboard *Mir* Was Dream Mission for Polyakov." RocketStem.com, 9 February 2015. www.rocketstem.org/2015/02/09/russian-cosmonaut-valeri-polyakov-spent-record-breaking-14-months-aboard-mir-space-station-in-1990s/#:~:text=%E2%80%9CWe%20can%20fly%20to%20Mars,traveled%20nearly%20187%20million%20miles.

Hall, Shannon. "Hubble Telescope Faces Threat from SpaceX and Other Companies' Satellites." *New York Times*, 2 March 2023. www.nytimes.com/2023/03/02/science/hubble-spacex-starlink.html?utm_source=newsletter&utm_medium=email&utm_campaign=newsletter_axiosspace&stream=science.

Harrigan, Stephen. "Mr. Hannah's Rocket." *Texas Monthly*, November 1982.

Hollingham, Richard. "The NASA Mission That Broadcast to a Billion People." BBC.com, 21 December 2018. www.bbc.com/future/article/20181220-the-nasa-mission-that-broadcast-to-a-billion-people#:~:text=The%20Apollo%20programme%20was%20on,a%20Christmas%20message%20from%20orbit.&text=It%20is%2021%20December%201968,am%2C%20Cape%20Kennedy%2C%20Florida.

Holmes, Steven A. "U.S. and Russians Join in New Plan for Space Station." *New York Times*, 3 September 1993.

Homans, Charles. "Edgar Mitchell." *New York Times Magazine*, 21 December 2016.

Howell, Elizabeth. "International Space Station: Facts about the Orbital Laboratory." Space.com, 23 August 2022. www.space.com/16748-international-space-station.html.

Huntsville (AL) Times. "Man Enters Space." 12 April 1961.

Ignatius, David. "The Space Force Needs to Get Bigger." *Washington Post*, 22 August 2023. www.washingtonpost.com/opinions/2023/08/22/us-space-force-military-pentagon-competition/.

Irwin, Don. "$1 Million Far-Out Tourist Flights Seen." *Los Angeles Times*, 29 April 1985.

Ivins, Molly. "Ed Who?" *New York Times Magazine*, 30 June 1974.

Kokonos, Lance, and Ian Ona Johnson. "The Forgotten Rocketeers: German Scientists in the Soviet Union, 1945–1959." *War on the Rocks*, 28 October 2019. https://warontherocks.com/2019/10/the-forgotten-rocketeers-german-scientists-in-the-soviet-union-1945-1959/.

Lagnado, Lucette. "A Scientist's Nazi-Era Past Haunts Prestigious Space Medicine Prize." *Wall Street Journal*, 30 November 2012.

SOURCES | 257

Lardner, George, Jr., and Walter Pincus. "Military Had Plan to Blame Cuba if Glenn's Space Mission Failed." *Washington Post*, 18 November 1997.

Larson, Peter. "Shuttle Inventor Sees Some Gear as Obsolete Now." *Orlando Sentinel*, 5 July 1982.

Lea, Robert. "Europe Wants to Build a Nuclear Rocket for Deep Space Exploration." Space.com, 3 May 2023. www.space.com/european-space-agency-nuclear -propulsion.

Lee, Michele Ye Hee, and Lily Kuo. "For Rivals Japan and China, the New Space Is about Removing Junk." *Washington Post*, 20 November 2022.

LePage, Andrew L. "Old Reliable: The Story of the Redstone." *Space Review*, 2 May 2011. www.thespacereview.com/article/1836/1.

Levy, Max G. "Scientists Discovered Exposed Bacteria Can Survive in Space for Years." *Smithsonian Magazine*, 26 August 2020.

Liebrum, Martha. "The Astronaut's Widow." *Houston Post*, 5 November 1974.

Lipton, Eric. "U.S. Is Increasing Ability to Fight Threats in Space." *New York Times*, 19 May 2024.

Loeb, Abraham. "Can the Universe Provide Us with the Meaning of Life?" *Scientific American*, 21 January 2020. https://blogs.scientificamerican.com/observations /can-the-universe-provide-us-with-the-meaning-of-life/.

Los Angeles Times. "Ex-Astronaut Warns as Dancers Chant." 15 April 1979.

Lyons, Richard D. "Senate Votes Aid to Space Shuttle." *New York Times*, 12 May 1972. www.nytimes.com/1972/05/12/archives/senate-votes-aid-to-space-shuttle -200million-is-authorized-for.html.

Maidenberg, Micah. "Elon Musk's SpaceX Now Has a De Facto Monopoly on Rocket Launches." *Wall Street Journal*, 7 July 2023.

Moore, Gary. "Firms Vying with NASA on Space Launches." *Atlanta Journal*, 31 July 1984.

Myers, Steven Lee. "China and Russia Agree to Explore the Moon Together." *New York Times*, 10 March 2021.

Neufeld, Michael. "Spinning out of Control: Gemini VIII's Near-Disaster." *Air & Space Quarterly*, 16 March 2021. https://airandspace.si.edu.

New York Daily News. "Irony: $1 Million to Rocket Pioneer Who Died Too Soon." 7 August 1960.

New York Times. "Titan Rocket with Secret Cargo Explodes." 3 August 1993.

O'Brien, Miles. "We Aimed for the Stars . . . Until We Stopped." *Space News*, 22 January 2009. https://spacenews.com/oped-we-aimed-starsuntil-we-stopped/.

O'Leary, Brian. "Rebellion among the Astronauts." *Ladies' Home Journal*, March 1970.

258 | SOURCES

O'Toole, Thomas. "More Astronauts Quit as Training and Travel Strain Family Lives." *Washington Post*, 28 October 1985.

Perez, Sarah. "Google Earth and Maps Get Sharper Images with Satellite Updates." Tech Crunch.com, 27 June 2016. https://techcrunch.com/2016/06/27/google-earth-and-maps-get-sharper-satellite-imagery-with-new-update/.

Perlez, Jane, and Grace Tatter. "Shared Secrets: How the U.S. and China Worked Together to Spy on the Soviet Union." WBUR.org, 18 February 2022. www.wbur.org/hereandnow/2022/02/18/great-wager-spy-soviet-union.

Pope, Charles. "Kendall Presents Unsparing Blueprint for Confronting China, Other Threats." United States Air Force website, 20 September 2021. www.af.mil/News/Article-Display/Article/2781521/kendall-presents-unsparing-blueprint-for-confronting-china-other-threats/.

Portree, David S. F. "Who Controls the Moon Controls the Earth (1958)." *Wired*, 31 March 2012. www.wired.com/2012/03/who-controls-the-moon-controls-the-earth-1958/.

Rivero, Nicolas. "A New Watchdog Satellite Will Sniff Out Methane Emissions from Space." *Washington Post*, 4 March 2024.

Rogin, Josh. "The Most Shocking Intel Leak Reveals New Chinese Military Advances." *Washington Post*, 13 April 2023. /www.washingtonpost.com/opinions/2023/04/13/china-hypersonic-missile-intelligence-leak/.

Roulette, Joey, and Marisa Taylor. "Musk's SpaceX Is Building Spy Satellite Network for U.S. Intelligence Agency." Reuters.com, 16 March 2024.

Scoles, Sarah. "The Doctor from Nazi Germany and the Search for Life on Mars." *New York Times*, 24 July 2020. www.nytimes.com/2020/07/24/science/mars-jars-strughold.html.

———. "Why We'll Never Live in Space." *Scientific American*, October 2023.

Spacewatch. "Low-Earth Orbit Debris Cluttering Space." November 1990.

Thomson, Jess. "China Uses Drag Sail to Clean Up Space Junk Successfully." *Newsweek*, 22 July 2022. www.newsweek.com/chinese-space-sail-removes-space-junk-orbit-1722670.

Trento, Joseph J., and Susan Trento. "Power Failure." *Regardie's Magazine*, April 1987.

Tumlinson, Rick. "Return to the Moon: The Race We Have to Win (Again)." Space.com, 20 July 2023. www.space.com/return-to-moon-china-space-race.

Twiss, Shannon. "The Environmental Impacts of the New Space Race." *Georgetown Environmental Law Review*, 7 April 2022. www.law.georgetown.edu/environmental-law-review/blog/the-environmental-impacts-of-the-new-space-race/.

Verrengia, Joseph B. "Astronauts: Weightless or Useless?" *Rocky Mountain News*, 14 April 1991.

Von Hippel, Frank, and Thomas B. Cochran. "The Myth of the Soviet 'Killer' Laser." *New York Times*, 19 August 1989.

West, Julian. "Shuttle Has Challenger." *The Province* (Vancouver BC), 23 October 1984.

Zatuchni, Stephen B. "What Happened to the Dream? Reagan's Plan Would Turn Space into a Factory." *Houston Post*, 17 February 1984.

Interviews and Personal Communications

Giles, Dave, and Emily Carney. "Being in Mission Control for Skylab: An Interview with Bill Moon." *Space and Things*, episode 142, 18 May 2023.

———. "An Interview with Rusty Schweickart: Apollo 9, Skylab, Asteroids, and So Much More . . ." *Space and Things*, episode 132, 9 March 2023.

Griffin, Michael D. "What about the NASA Appropriations Cut." Email, 2 August 2004.

Kerwin, Dr. Joseph P. Interview by Kevin Rusnak, 12 May 2000. NASA Johnson Space Center Oral History Project. https://historycollection.jsc.nasa.gov /JSCHistoryPortal/history/oral_histories/KerwinJP/KerwinJP_5-12-00.htm.

McCandless, Bruce, II. Letter to Mr. Norm Augustine, Chairman, Review of U.S. Human Space Flight Plans Committee. 17 September 2009.

Moore, Patrick. Interview of Neil Armstrong. "The Sky at Night," BBC Television, 1970. https://www.youtube.com/watch?v=EIPn_iuLPA4.

Pyle, Rod, and Tariq Malik. Interview of Dr. Ed Lu. *This Week in Space*, episode 67, 1 July 2023.

———. Interview of Dr. Franklin Chang-Díaz. *This Week in Space*, episode 80, 22 September 2023.

Russell, Matthew. Interview of Jaime Green. *The Interplanetary Podcast*, episode 293, 17 May 2023.

Other Sources

"American Rocketeer." NASA Jet Propulsion Laboratory, 11 January 2021. www .jpl.nasa.gov/who-we-are/documentary-series-jpl-and-the-space-age/episode-1.

"Biggest Asteroid Impacts in Earth's History." World Atlas.com. www.worldatlas .com/articles/biggest-asteroid-impacts-in-earth-s-history.html.SP-4012.

Chapman, Phil. "Icebergs Ahead." Memorandum to Scientist-Astronauts, 6 August 1970.

Crippen, Robert L. "Space Shuttle Title." Memorandum for General Distribution, 22 February 1990.

Dell'Osso, Ray. "EVA." Informal Memorandum to CG3/Crew Systems Section Personnel, 27 October 1982.

260 | SOURCES

Global Climate Change Fact Sheet. NASA Science website. Retrieved 27 September 2020. https://science.nasa.gov/climate-change/evidence/.

"How a Gyroscope Guides a Rocket." V-2 Rocket History.com, 20 April 2018. https://v2rockethistory.com/gyroscope-guides-rocket/.

"How Do We Know Climate Change Is Real?" NASA Science website. Retrieved 27 September 2020. https://science.nasa.gov/climate-change/.

"LandSat Satellite Missions." United States Geological Survey website. https://www.usgs.gov/landsat-missions/landsat-satellite-missions.

"Lunar Rocks and Soils from Apollo Missions." NASA Curation website. https://curator.jsc.nasa.gov/lunar/#:~:text=Between%201969%20and%201972%20six,exploration%20sites%20on%20the%20Moon.

"The Nancy Grace Roman Space Telescope." NASA Jet Propulsion Laboratory website. Retrieved 24 February 2022. www.jpl.nasa.gov/missions/the-nancy-grace-roman-space-telescope.

"NASA, DARPA Will Test Nuclear Engine for Future Mars Missions." NASA website. Press Release, Release 23-012, 24 January 2023. www.nasa.gov/press-release/nasa-darpa-will-test-nuclear-engine-for-future-mars-missions.

NASA Historical Data Book: Volume IV, NASA Resources 1969–1978, Chapter Three: NASA Personnel. https://history.nasa.gov/SP-4012/vol4/ch3.htm.

"NASA Identifies Unidentified Anomalous Phenomena Study Team Members." NASA website, 21 October 2022 (updated 22 December 2022). www.nasa.gov/feature/nasa-announces-unidentified-anomalous-phenomena-study-team-members/.

Puddy, Donald R. "Utilization of the Extravehicular Activity (EVA) Resource." Memorandum to Manager, Space Station Projects Office, CB-88-246, 23 October 1988.

"The Rocket Men." NASA Jet Propulsion Laboratory, 17 January 2007. www.youtube.com/watch?v=FhDKjMebFWU.

Stafford, Thomas P. "Astronaut Group Prestige and Morale." Memorandum to All Astronauts, 8 October 1969.

In the Outward Odyssey: A People's History of Spaceflight series

Into That Silent Sea: Trailblazers of the Space Era, 1961–1965
Francis French and Colin Burgess
Foreword by Paul Haney

In the Shadow of the Moon: A Challenging Journey to Tranquility, 1965–1969
Francis French and Colin Burgess
Foreword by Walter Cunningham

To a Distant Day: The Rocket Pioneers
Chris Gainor
Foreword by Alfred Worden

Homesteading Space: The Skylab Story
David Hitt, Owen Garriott, and Joe Kerwin
Foreword by Homer Hickam

Ambassadors from Earth: Pioneering Explorations with Unmanned Spacecraft
Jay Gallentine

Footprints in the Dust: The Epic Voyages of Apollo, 1969–1975
Edited by Colin Burgess
Foreword by Richard F. Gordon

Realizing Tomorrow: The Path to Private Spaceflight
Chris Dubbs and Emeline Paat-Dahlstrom
Foreword by Charles D. Walker

The X-15 Rocket Plane: Flying the First Wings into Space
Michelle Evans
Foreword by Joe H. Engle

Wheels Stop: The Tragedies and Triumphs of the Space Shuttle Program, 1986–2011
Rick Houston
Foreword by Jerry Ross

Bold They Rise: The Space Shuttle Early Years, 1972–1986
David Hitt and Heather R. Smith
Foreword by Bob Crippen

Go, Flight! The Unsung Heroes of Mission Control, 1965–1992
Rick Houston and Milt Heflin
Foreword by John Aaron

Infinity Beckoned: Adventuring Through the Inner Solar System, 1969–1989
Jay Gallentine
Foreword by Bobak Ferdowsi

Fallen Astronauts: Heroes Who Died Reaching for the Moon, Revised Edition
Colin Burgess and Kate Doolan with Bert Vis
Foreword by Eugene A. Cernan

Apollo Pilot: The Memoir of Astronaut Donn Eisele
Donn Eisele
Edited and with a foreword by Francis French
Afterword by Susie Eisele Black

Outposts on the Frontier: A Fifty-Year History of Space Stations
Jay Chladek
Foreword by Clayton C. Anderson

Come Fly with Us: NASA's Payload Specialist Program
Melvin Croft and John Youskauskas
Foreword by Don Thomas

Shattered Dreams: The Lost and Canceled Space Missions
Colin Burgess
Foreword by Don Thomas

The Ultimate Engineer: The Remarkable Life of NASA's Visionary Leader George M. Low
Richard Jurek
Foreword by Gerald D. Griffin

Beyond Blue Skies: The Rocket Plane Programs That Led to the Space Age
Chris Petty
Foreword by Dennis R. Jenkins

A Long Voyage to the Moon: The Life of Naval Aviator and Apollo 17 Astronaut Ron Evans
Geoffrey Bowman
Foreword by Jack Lousma

The Light of Earth: Reflections on a Life in Space
Al Worden with Francis French
Foreword by Dee O'Hara

Son of Apollo: The Adventures of a Boy Whose Father Went to the Moon
Christopher A. Roosa
Foreword by Jim Lovell

Star Bound: A Beginner's Guide to the American Space Program, from Goddard's Rockets to Goldilocks Planets and Everything in Between
Emily Carney and Bruce McCandless III

To order or obtain more information on these or other University of Nebraska Press titles, visit nebraskapress.unl.edu.